SCIENTIFIC PERSPECTIVISM

SCIENTIFIC PERSPECTIVISM

RONALD N. GIERE

The University of Chicago Press : CHICAGO AND LONDON

The University of Chicago Press, Chicago 60637
The University of Chicago Press, Ltd., London
© 2006 by The University of Chicago
All rights reserved. Published 2006
Paperback edition 2010
Printed in the United States of America

19 18 17 16 15 14 13 12 11 10   2 3 4 5 6

ISBN-13: 978-0-226-29212-0 (cloth)
ISBN-13: 978-0-226-29213-7 (paper)
ISBN-10: 0-226-29212-6 (cloth)
ISBN-10: 0-226-29213-4 (paper)

Library of Congress Cataloging-in-Publication Data

Giere, Ronald N.
    Scientific perspectivism / Ronald N. Giere.
        p.   cm.
    Includes bibliographical references and index.
    ISBN 0-226-29212-6 (cloth : alk. paper)
    1. Science—Philosophy.   2. Science—History.   I. Title.

Q175.G48899 2006
501—dc22

                                    2006002489

CONTENTS

List of Illustrations vii

Acknowledgments ix

1. SCIENTIFIC KNOWLEDGE 1
What Is the Problem? • Objective Realism • Constructivism •
Naturalism • Perspectivism • Retrospect

2. COLOR VISION 17
Introduction • Basic Color Science • Color Subjectivism •
Color Objectivism • Comparative Color Vision • Color Perspectivism •
The Philosophy of Color • A Final Question

3. SCIENTIFIC OBSERVING 41
Introduction • Astronomy in Color • Deep Space from the
Perspective of the Hubble Telescope • The Milky Way in Gamma
Ray Perspectives • Conclusions within Perspectives • Imaging
the Brain • Instrumental Perspectives

4. SCIENTIFIC THEORIZING 59
Introduction • Representing • Theories • Laws of Nature •
Fitness • Maps • Truth within a Perspective • Perspectives and
Paradigms • Scientific Kinds • Perspectival Realism •
The Contingency Thesis Revisited

5. PERSPECTIVAL KNOWLEDGE
AND DISTRIBUTED COGNITION 96
Introduction • Distributed Cognition • Scientific Observation
as Distributed Cognition • Models as Parts of Distributed Cognitive
Systems • Computation in Scientific Distributed Cognitive Systems •
Agency in Scientific Distributed Cognitive Systems • Why Distributed
Cognition? • Distributed Cognition and Perspectival Knowledge

Color gallery follows page 62

Notes 117

References 137

Index 147

ILLUSTRATIONS

CHAPTER TWO

2.1 Relative spectral sensitivity of color pigments for trichromats.
2.2 Neural code for processing pigment activation in trichromats.
2.3 Chromatic-response function for trichromats.
2.4 Wavelength discrimination as a function of wavelength and intensity.
2.5 Metamers for pure yellow.
2.6 Surface spectral reflectances for several common surfaces.
2.7 Metameric surface spectral reflectances.
2.8 Relative spectral sensitivity of color pigments for dichromats.

CHAPTER THREE

3.1 Path of the signal from deep space to the Space Telescope Science Institute.
3.2 Diagram of the COMPTEL instrument.
3.3 Chronology and resolution of imaging technologies.
3.4 Design of instrument for CAT.
3.5 The physics behind PET.
3.6 A schematic representation of a PET machine.
3.7 The longitudinal NMR field.
3.8 Localizing the source of the emitted radio frequency signal.

CHAPTER FOUR

4.1 Overview of a model-based account of theories.
4.2 A partial tourist map of central Washington, D.C.
4.3 A partial subway map of Washington, D.C.
4.4 A medieval T-O map.
4.5 A line drawing used to represent parts of North America.
4.6 Three projections of the Earth's surface onto two dimensions: Mercator, Peters, and Robinson.
4.7 Mechanical kinds generated by the principles of classical mechanics.
4.8 Schematic rendering of an experimental test of model fit.

CHAPTER FIVE

5.1 The multiplication of two three-digit numbers using a traditional technique.

5.2 Diagrams used to prove the Pythagorean Theorem.

5.3 Diagrammatic reasoning in standard artificial intelligence.

5.4 Diagrammatic reasoning as distributed cognition.

5.5 Bad versus good design of diagrams, based on cognitive principles.

PLATES

1 The electromagnetic spectrum.

2 The hue circle.

3 Comparison of trichromatic and dichromatic views of the same scene.

4 The Trifid Nebula.

5 Hubble (ACS) photo of January 2003.

6 COMPTEL image of the center of the Milky Way.

7 OSSE image of the center of the Milky Way.

8 Color photograph of a stained axial slice of human brain.

9 A representative CAT scan.

10 A representative PET scan.

11 A representative MRI image.

12 A representative fMRI image.

ACKNOWLEDGMENTS

I began thinking about the possibility of a perspectival realism as a development of constructive realism in the mid-1990s. This possibility was announced in the conclusion of *Science without Laws* (1999). The present book is an attempt to fulfill the promise of that announcement. I had the honor of presenting a first draft as the Leibniz Lectures under the auspices of the University of Hanover in the reconstructed Leibniz Haus in June 1999. I thank Paul Hoyningen-Huene for the invitation and his associates at the time, Marcel Weber and Howard Sankey, for a most enjoyable occasion. Among others who have supported my efforts, I count especially Paul Teller, Fred Suppe, Bas van Fraassen, and Arthur Fine. My Minnesota colleagues, Helen Longino, Wade Savage, and Ken Waters, provided much appreciated moral and professional support. I have also benefited from discussions with colleagues at universities in Great Britain, Finland, and The Netherlands, as well as at home in the United States. The National Science Foundation, the National Endowment for the Humanities, the National Humanities Center, and The Netherlands Institute for Advanced Study have provided institutional support. I can only hope that the result proves worthy of their support.

CHAPTER ONE
SCIENTIFIC KNOWLEDGE

What Is the Problem?

Since the end of World War II, the influence of science within Western culture has come to rival that of other major components of culture, such as religion, the military, law, medicine, entertainment, and even commerce. Indeed, many other components of Western culture have been largely transformed by technological developments based on new scientific knowledge, medicine and the military being prime examples. If one includes computer technology, one could argue that virtually all aspects of Western culture are being transformed with the help of new scientific knowledge.[1]

In spite of the undeniable successes of post–World War II science, there have always been some who questioned whether the influence of science within Western society is a uniformly good thing. And with reason. The bomb that won the war in the Pacific became the basis for the doctrine of mutually assured destruction, which governed the long Cold War that followed. Yet even those who raised such questions typically ascribed the problems to the *applications* of scientific knowledge. The nature of the knowledge itself was rarely questioned. It was taken for granted that scientists were discovering the objectively real inner workings of nature. These workings are there to be discovered. It only takes effort, sometimes requiring huge expenditures of resources, to uncover them. This pretty well sums up the attitude of most people at the beginning of the twenty-first century, including both scientists and nonscientists.

After World War II, widespread serious questions about the nature of scientific knowledge itself began to be raised only in the mid-1960s to mid-1970s. During this period, many who grew up after World War II found themselves horrified by the use of B-52s and other high-powered military technology against Vietnamese peasants riding bicycles and armed with little more powerful than an AK-47. Around the same time, it became clear that modern industries and, particularly, agricultural technologies were degrading the environment in many ways.[2] Additionally, some women began to regard new household technologies as more enslaving than liberating.[3] Some people whose fundamental attitudes were formed during this period became university professors. And a few of these focused their attention on the sciences, not as scientists themselves, but as critics of science. Not surprisingly, many of these critics found their way into the humanities and social sciences, such as history, philosophy, sociology, literature, or, eventually, cultural studies. Especially in the latter three disciplines, it is now almost taken for granted that scientific knowledge is some sort of "social construct." I will shortly try to make clearer what this might mean. Here, it is enough to know that scholars in these disciplines typically deny that scientific activity amounts to the discovery of something unproblematically out there in the world waiting to be uncovered by clever scientists. Rather, these constructivist scholars would argue, there is at best a consensus among scientists regarding what to *say* they have found. And reaching a consensus is a complex social process in which exhibiting empirical evidence is only a part, and by no means a determining part. Evidence, it is claimed, must be interpreted, and this can always be done in more than one way.

So now we have a major disagreement within the academic world on something as substantial and important as how to understand the nature of claims to scientific knowledge. Most scientists and representatives of the general public say one thing. Many scholars in the humanities and social sciences, including most in the fledging field of science studies, say something quite different. Nor has this dispute been entirely academic. At the end of the twentieth century, it broke into the public media in the United States and Britain as "The Science Wars." As in most wars, there was considerable collateral damage. Positions became polarized. Careers were altered. In the end, the most extreme advocates of constructivist views were chastened, abandoned by their more sober colleagues. Most scientists who were exposed to the conflict seem to have remained comfortable with their original views. It is difficult to discern any reaction, one way or another, among members of the general public.[4]

Among humanists, philosophers of science have traditionally been closest to sharing the views of scientists themselves. The project of Anglo-American philosophers of science, in particular, has generally been to analyze both the theories and methods of scientists with the aim of making explicit what remains implicit in what scientists do. To be sure, this analysis often involves a good deal of interpretation and reconstruction, whether focused on particular sciences or on science more generally. One major dispute among these philosophers of science has been whether all the claims of scientists about entities and processes described in scientific theories should be interpreted realistically or in some less committal manner. This dispute to some extent intersects with the broader dispute over constructivist interpretations of scientific claims, because the less committal the scientific claims, the more room there is for constructivist interpretations.

I can now give a preliminary statement of my objective in writing this book. It is to develop an understanding of scientific claims that mediates between the strong objectivism of most scientists, or the hard realism of many philosophers of science, and the constructivism found largely among historians and sociologists of science. My understanding will turn out to be closer to that of objectivist scientists and realist philosophers of science than to that of constructivists. Nevertheless, in spite of the excesses of some constructivists, I think there is a valid point to the constructivist critique of science. The challenge is to do justice to this point while avoiding the excesses.

There already exists a considerable literature containing arguments for or against realism in the philosophy of science and another somewhat over-lapping literature with arguments for or against constructivism in science studies.[5] When I think it productive, I will engage some of this literature, although often in footnotes rather than in the body of the text. Both philosophers and sociologists sometimes complain that the debate between realists and empiricists in the philosophy of science, or between realists and constructivists, has been inconclusive, even sterile. To some extent I share this opinion. Rather than simply abandoning or ignoring the issues, how-ever, I seek to change the terms of the debate by developing an alternative view that is more than a minor variant on already existing views.

The positive view to be developed here is a version of *perspectivism*. Perspectivism has antecedents in the work of some much earlier philoso-phers, such as Leibniz, Kant, and Nietzsche.[6] My perspectivism, however, will be developed almost wholly within the framework of contemporary sci-ence. Thus, in the end, my own claims must be reflexively understood as themselves perspectival. The remainder of this introductory chapter will be devoted to further clarifying both objectivism and constructivism and then to introducing my own version of perspectivism. I would caution the reader not to jump to hasty conclusions as to my ultimate understanding of perspectivism. This will emerge in the chapters to come.

Objective Realism

Everyone starts out a common-sense realist. Among the first things a child learns is to distinguish itself from the things around it. Pretty soon the independent reality of ordinary things—trees, dogs, other people—is taken for granted. Things are thought to be just what they seem to be. For most people most of the time, common-sense realism works just fine.[7]

The realism of scientists may be thought of as a more sophisticated version of common-sense realism. Here are several expressions of objectivist realism by a Nobel Prize–winning physicist, Steven Weinberg (2001).[8] The first emphasizes the discovery of truths and the permanence of scientific knowledge.

> What drives us onward in the work of science is precisely the sense that there are truths out there to be discovered, truths that once discovered will form a permanent part of human knowledge. (126)

The second adds the idea that the truths to be discovered take the form of laws.

> [A]side from inessentials like the mathematical notation we use, the laws of physics as we understand them are nothing but a description of reality. (123)

The third emphasizes a sense of progress toward truths.

> I can't see any sense in which the increase in scope and accuracy of the hard parts of our theories is not a cumulative approach to truth. (126)

These expressions of the nature of scientific claims are more explicit than one usually finds in the writings of scientists, because here Weinberg is deliberately acting in the role of spokesperson for the scientific community. There is no reason to think, however, that he does not represent the sentiments of a great many scientists.[9]

The themes expressed by Weinberg may also be found in the writings of many philosophical scientific realists.[10] They coincide with a viewpoint the philosopher Hilary Putnam a generation ago called "metaphysical realism," namely, "There is exactly one true and complete description of 'the way the world is.'" (1981, 49) Putnam argued that metaphysical realism is ultimately an incoherent doctrine, and recommended instead what he called "internal realism," a view sometimes characterized as perspectival (Blackburn 1994). I disagree with aspects of Putnam's analysis, but it is not primarily because of these disagreements that I prefer different labels. Rather, it is because the logical-linguistic framework in terms of which Putnam frames his presentation is shared by almost no one in the scientific or the science studies community apart from analytic philosophers. His work is not so much even philosophy of science as it is logic, philosophy of language and mind, epistemology, and metaphysics.[11] This work thus fails to engage many of those whom I wish to engage. For this reason, I have chosen the term "objectivist realism" for views like those expressed by Weinberg and in the language in which he expressed them. My rejection of objectivist realism is based entirely on an examination of scientific practice, something appreciated by scientists as well as historians, sociologists, psychologists, and other students of science as a human enterprise. Nevertheless, some further elaboration is required.

In the passages quoted, and in other writings, Weinberg seems to be saying not only that there are true laws to be discovered, but that scientists are capable of discovering them and, moreover, knowing that they have discovered them, so that, in his words, these truths "once discovered will form a permanent part of human knowledge." In spite of the epistemological optimism of these words, I doubt Weinberg would deliberately claim for scientists the kind of infallibility he implies. In writings too long or scattered to quote here, he seems to agree with most contemporary philosophical opinion that no empirical claims can be known with absolute certainty to be true. It always remains possible we will find out later that we were earlier mistaken.[12] By itself, however, this admission does not imply that scientists are not, in some weaker sense, fully justified in making unqualified claims to knowledge of the objective truth of some laws of nature. The way to resolve this apparent tension is to take objective realism as an expression the proper *aim* of scientific investigation, even though it can never be known with certainty that the aim has been achieved. This position has been succinctly stated by the foremost antirealist philosopher of science of the later twentieth century, Bas van Fraassen. "The correct statement of scientific realism," he writes, is: "Science aims to give us, in its theories, a literally true story of what the world is like; and acceptance of a scientific theory involves the belief that it is true" (1980, 8). I take this to be a good capsule statement of objectivist realism, but I do not want to identify scientific realism in general with objective realism. I will be arguing that there is a kind of realism that applies to scientific claims that is more limited than this full-blown objective realism. Thus, in the end, I wish to reject objective realism but still maintain a kind of realism, a perspectival realism, which I think better characterizes realism in science.[13] For a perspectival realist, the strongest claims a scientist can legitimately make are of a qualified, conditional form:

"According to this highly confirmed theory (or reliable instrument), the world seems to be roughly such and such." There is no way legitimately to take the further objectivist step and declare unconditionally: "This theory (or instrument) provides us with a complete and literally correct picture of the world itself."[14]

The main thrust of the arguments to be presented in this book is to show that the practice of science itself supports a perspectival rather than an objectivist understanding of scientific realism. This does not, however, constitute an argument for constructivism in general, but only for the position that scientific claims may be in part socially constructed, and thus for the *possibility* of discovering the social contributions to these claims. The extent to which any particular scientific claim is socially constructed can only be determined, if at all, by a detailed historical examination of the case in question.

Constructivism

The first book containing the expression "social construction" with its contemporary meaning in the title was Berger and Luckmann's *The Social Construction of Reality*, published in 1966.[15] In fact, the subject of Berger and Luckmann's book is the social construction of *social* reality. Hardly anyone would question that a large part of our social world is the product of human social interaction over time.[16] For example, that there are males and females might be regarded as a biological fact, that there are men and women might be ambiguously either a biological fact or a social fact, but that there are husbands and wives (as opposed to just mates) is unambiguously a social fact. That there could be anything that is a husband or a wife requires that there be an institution of marriage. And that there be such an institution requires a fairly complex form of social organization.[17] So, sometime early in the history of humans there were males and females but no husbands and wives, because there was no institution of marriage. This was so even if males and females typically bonded into pairs. In general, then, the possibility of there being members of a social category depends on the existence of a social arrangement in which that category makes sense.

Unlike many later constructivists, Berger and Luckmann clearly distinguished between biological and social facts, declaring that any attempt to legislate that men should bear children "would founder on the hard facts of human biology" (1966, 112). Now consider a seemingly unproblematic example from the physical sciences. Before the Scientific Revolution, what we now call planets were known as wandering stars. Indeed, the word *planet* derives from a Greek word meaning "wanderer." So the category of "planet" with our modern meaning did not exist prior to the seventeenth century. Are we to conclude that planets did not exist before the seventeenth century? That sounds absurd to the modern scientific ear. Of course they existed before then; indeed, according to currently accepted theories, they existed a few billion years before then. As Weinberg, ever the objective realist, puts it: "[I]t is true that natural selection was working during the time of Lamarck, and the atom did exist in the days of Mach, and fast electrons behaved according to the laws of relativity even before Einstein" (2001, 120). I will recommend a more modest way of maintaining such claims that now seem pretty much common sense.

The Contingency Thesis

Lest Weinberg be accused of merely stating the obvious when claiming that "the atom did exist in the days of Mach," several founders of the constructivist movement did early on make claims that suggested the contrary. Bruno Latour and Steve Woolgar (1979, 128–29), for example, explicitly rejected the idea that objects of investigation have "an independent existence 'out there,' " insisting, rather, that these objects are "constructed" in the sense of being "constituted solely through the use of" various recording devices. Similarly, Karin Knorr-Cetina (1983, 135) spoke of "scientific reality as progressively emerging out of indeterminacy and (self-referential) constructive operations, without assuming it to match any pre-existing order of the real."[18]

A more moderate version of constructivism was advocated by the members of the now mostly dispersed Edinburgh School under the banner "Sociology of Scientific Knowledge" (SSK). As I now understand them, these constructivists did not deny that the world has a definite "objective" structure or argue that reality is in some sense "constituted" by human activities. Rather, they argued that the process of doing science is so infused with all sorts of human judgments and values that what ends up being *proclaimed* to be the structure of reality may bear little resemblance to the real structure of the world. Members and associates of the Edinburgh school issued a number of theoretical proclamations, including David Bloor's (1976) "Strong Programme in the Sociology of Scientific Knowledge" and Harry Collins's (1981) "Empirical Program of Relativism." The real strength of the program, however, lay in its case studies, both historical and contemporary (Edge and Mulkay 1976; MacKenzie 1981). Early among these were Steve Shapin's (1975, 1979) studies of phrenology in late-eighteenth-century Edinburgh. In these studies Shapin attempted to show that support for phrenology was concentrated among members of "rising bourgeois groups," while opposition was strong within "traditional elites." Investigators, he concluded, saw what it was in their social interest to see. Summing up, he wrote: "Reality seems capable of sustaining more than one account given of it, depending upon the goals of those who engage with it; and in this instance at least those goals included considerations in the wider society such as the redistribution of rights and resources among social classes" (1982, 194). Here we have an explicit statement of what I, following Hacking (1999), will call the *contingency thesis*. Shapin's version is: "Reality seems capable of sustaining more than one account of it." Here I will only note that the thesis itself is distinct from any account of how, in practice, contingency is overcome to yield a unique conclusion, which is at least the proclaimed result of most scientific investigations. In this case, Shapin claims it is class interests that do the trick. But this is only one of many possibilities, assuming the contingency thesis is correct, which, of course, is strongly contested.

Here is another early version of the contingency thesis from the writings of Harry Collins, initially an ally of the Edinburgh School and long associated with the University of Bath: "[I]n one set of social circumstances 'correct scientific method' applied to a problem would precipitate result p whereas in another set of social circumstances 'correct scientific method' applied to the same problem would precipitate result q, where, perhaps, q implies not-p" (1981, 6–7). Here Collins adds the idea that even the application of "correct scientific method" is not sufficient to eliminate the contingency in which result gets accepted.

Collins's early work focused on contemporary research aimed at detecting gravity waves.[19] This suggests an explanation for why both Shapin and Collins found contingency in science. They were dealing with cases in which the science is weak, phrenology having turned out to be mostly worthless and the detection of gravity waves being at the limits of experimental capabilities. If the science is weak, there is correspondingly more opportunity for interests of all kinds to influence what different investigators might conclude.

This explanation, however, does not apply to one of the most extensive works to come out of Edinburgh, Andy Pickering's *Constructing Quarks: A Sociological History of Particle Physics* (1984). In this book, Pickering details the transition from the "old" high-energy physics (HEP) of the 1960s to the "new" HEP of the late 1970s. Pickering contrasts what he calls the "scientist's account" of this history with his own "social" account. On the scientist's account, the transition from the old HEP to the new was driven by experimental results in the context of new theoretical ideas. Pickering claims that the scientist's account is really a retrospective rationalization based on the independent judgment that the new HEP provides the better account of reality. On Pickering's own account, experiments were not decisive. Rather, interpreting an experimental result requires professional judgment that is subject to many influences. In particular, judgment is subject to what he calls "opportunism in context." Scientists interpret results partly in terms of their own expertise, the available instrumentation, and their judgment as to which approaches provide the most opportunity for doing new work, whether experimental or theoretical. For Pickering, then, adopting a new theoretical approach involves creating a new research tradition. And, finally, if the social and material conditions had been different, the conclusion could have been different. Physicists need not have created the new HEP tradition.[20]

In his most recent book, *The Mangle of Practice* (1995), Pickering provides a somewhat different account of contingency in science. The general picture is one of scientific research proceeding through a process of *resistance* and *accommodation* as scientists attempt to reach a stable configuration of theory and experiment through interaction with nature. It is this process he calls "the mangle of practice." Here the contingency thesis is expressed in terms of what Pickering calls historical "path dependence" (1995, 185).

> [M]y analysis of practice . . . points to a *situatedness* and *path dependence* of knowledge production. On the one hand, what counts as empirical or theoretical knowledge at any time is a function not just of how the world is but of the specific material-conceptual-disciplinary-social-etc. space in which knowledge production is situated. On the other hand, what counts as knowledge is not determined by the space in which it is produced. . . . [O]ne needs also to take into account the contingencies of practice, the precise route that practice takes through that space. The contingent tentative fixing of modeling vectors, the contingent resistances that arise, the contingent formulation of strategies of accommodation, the contingent success or failure of these— all of these structure practice and its products.

The general thesis is that contingency in the eventual knowledge claims is due to contingencies in the research process. The *product*, claims to knowledge, is determined in part by the *process* of research.

The late James Cushing's *Quantum Mechanics: Historical Contingency and the Copenhagen Hegemony* (1994) provides a final example from the heart of objective realist territory, fundamental physics. The claim of historical contingency in science present in the title is elaborated in the very first paragraph of the preface. "The central theme of this book is that historical contingency plays an essential and ineliminable role in the construction and selection of a successful scientific theory from among its observationally equivalent and unrefuted competitors. I argue that historical contingency, in the sense of the order in which events take place, can be an essential factor in determining which of two empirically adequate and fruitful, but observationally equivalent, scientific theories is accepted by the scientific community." The overall theme is that the continuing general rejection of the 1952 Bohm interpretation of quantum mechanics is due largely to the fact that the now-standard Copenhagen interpretation was already by 1950 so well entrenched. And Cushing was a professor of physics!

This case is obviously complex and requires considerable specialized expertise to examine in any detail.[21] I mention it mainly to provide another example of serious attempts to show contingency in the most prominent parts of fundamental physics. These attempts are, however, implicitly dismissed by objective realists such as Weinberg:

> I think that physical theories are like fixed points, toward which we are attracted. Our starting points may be culturally determined, our paths may be affected by our personal philosophy, but the fixed point is there nonetheless. It is something toward which any physical theory moves, and when we get there we know it, and then we stop. . . . The . . . physics . . . of fields and elementary particles . . . is moving toward a fixed point. But this fixed point is unlike any other in science. That final theory toward which we are moving will be a theory of unrestricted validity, a theory which is applicable to all phenomena throughout the universe, a theory which, when we discover it, will be a permanent part of our knowledge of the world. (2001, 126–27)

His fellow Nobel Prize winner, Sheldon Glashow, is even more emphatic, declaring: "We believe that the world is knowable: that there are simple rules governing the behavior of matter and the evolution of the universe. We affirm that there are eternal, objective, extra-historical, socially-neutral, external and universal truths. The assemblage of these truths is what we call Science, and the proof of our assertion lies in the pudding of its success" (1992, 27). Here Glashow is explicitly asserting that the process of doing science leaves no trace in the product, a body of "universal truths."[22] The contrast with Pickering, Cushing, and other moderate constructivists could not be stronger. It would not be too extreme to call Glashow an "absolute objectivist." It is this sort of absolute objectivism that I take to be refuted by the practice of science itself.

Reflexivity

In traditional philosophical terms, moderate constructivism is more like an epistemological position than a metaphysical position. Regarding the meaning of scientific statements and the ultimate constitution of the world, moderate constructivists seem to be at one with objective realists. Thus, according to moderate constructivism, one could, in principle, be an objective realist about the *goal* of scientific claims, but then go on to argue that, as a matter of empirical fact, given the nature of human beings and human society, this ideal is not attainable. The claims scientists actually are capable of making (sometimes? often? always?) owe more to their social interactions than to their interactions with any objective reality. This seems to be the intended lesson of the many case studies produced by constructivists.[23]

Many critics of constructivism have raised the *reflexive* question of whether their conclusions about the nature of scientific claims do not also apply to the claims of constructivists themselves.[24] If not, constructivists would be claiming for themselves an objectivity they deny to the sciences. It would be difficult to justify such a privileging of the sociology of science.[25] On the other hand, if their own work comes under the scope of their conclusions, that would seem to make constructivism self-defeating. What social interests, for example, lead constructivists to their conclusions about science generally?

There is a way out of this dilemma. The conclusion that constructivism is self-defeating assumes that constructivists intend to present the objective truth about science. But that may not be their goal. Some constructivists seem to have, or have had, the goal of *undermining* the claims of scientists to have discovered the objective truth about nature. If this is the goal, then it is merely an unavoidable irony that their own claims are similarly undermined. What matters is that others come to question the claims of scientists to have discovered objective truths about nature.[26] The price of taking this way out of the reflexive dilemma, of course, is aligning oneself with radical critics of science.

Naturalism

Before describing a positive alternative to either objectivist realism or constructivism, I need to make clear the general framework within which I will be working. The accepted name of the framework is *naturalism*. Minimally, naturalism implies the rejection of appeals to anything supernatural. In the present context, that should not be much of a restriction. More generally, naturalism also implies the rejection of appeals to a priori claims of any kind. Thus, all claims, however well grounded empirically, are regarded as fallible. I would like this to be understood as not contradicting claims by scientists such as Weinberg and Glashow who, in spite of their insistence on the permanence of some scientific claims, would, if challenged, be unlikely to uphold a doctrine of infallibility for science. Because constructivists are already naturalists, many explicitly so, we thus all may be taken to share at least a general naturalist framework in which to carry on further debate.[27]

In spite of this local agreement, some common misunderstandings about naturalism need to be addressed. First, to what extent can naturalism itself be justified? Is not the denial of supernatural forces as metaphysical a claim as their affirmation? And can the blanket rejection of claims to a priori knowledge be itself anything less than an a priori claim?

Second, how can one determine the boundary of the natural beyond which lies the supernatural? Here naturalists typically appeal to the findings of modern science, but there are severe problems with this response. One cannot take today's natural science as determining the boundary of what is natural, simply because that boundary keeps moving. Once life itself was considered beyond the province of natural science, requiring a supernatural source. Similar claims are even now sometimes made for aspects of human consciousness (Adams 1987). The naturalist confidence that consciousness will eventually be given a natural scientific explanation seems to beg the question against the supernaturalist.

Both of the above problems can be eliminated by taking naturalism not as a *doctrine* but as a *methodological stance*.[28] When confronted with a seemingly intractable phenomenon, the naturalist supports research intended to produce a natural scientific explanation. The naturalist hopes, even expects, that this research will eventually be successful. This stance can be justified, to the extent that it needs to be justified at all, simply by appeal to past successes. We have explained life. Why not consciousness?

Note that the strategy of replacing metaphysical doctrines by methodological stances is less appealing to a supernaturalist or apriorist. The naturalist can wait until success is achieved. And there are good naturalist standards for when this happens. One typically appeals to supernatural explanations or a priori principles because of a pressing need to resolve some issue. Few theists since Pascal have found what I would call "methodological theism" very satisfying.[29] Moreover, it is debatable whether there are equally good criteria for successful supernaturalist or apriorist projects.

Note also that methodological naturalism is not a weak position. The methodological naturalist is free to criticize arguments for metaphysical and apriorist claims. A naturalist would typically attempt to show that such arguments are question begging, lead to an infinite regress, or are in some other way unsound. This, in fact, makes methodological naturalism a quite strong position.

Finally, it is often claimed that naturalism necessarily leaves out the normative dimensions of science. This does not bother most constructivists, who refuse to judge any particular scientific work as "good" or "bad." It does bother objectivist realists, who typically want to say that some methods are better than others and some theories better justified than others. What matters here is the *basis* for normative judgments. Naturalists reject a priori or otherwise "conceptual" justifications in favor of *instrumental* justifications. Roughly speaking, a method is good to the extent that it tends to select hypotheses with desirable characteristics, such as agreement with data or wide applicability, over hypotheses that lack these characteristics. Hypotheses are justified to the extent that they are selected by good methods. These judgments are themselves empirical and thus acceptable to naturalists, who require no further justification.

Of course, judging the reliability of a particular method requires applying some other method. Many have tried to avoid the threatening regress of methods by seeking a method that requires no further method for its justification, a "foundational" method. Naturalists regard such a search as hopeless and thus embrace some aspects of Pragmatism. The Pragmatist stance in this regard is to allow any particular method to be questioned, but not all methods at once. I will not here argue further for this particular Pragmatist doctrine, but rest content with noting its connection with a naturalist stance.[30]

Perspectivism

In common parlance, a perspective is often just a point of view in the sense that, on any topic, different people can be expected to have different points of view. This understanding is usually harmless enough in everyday life, but it can be pushed to the absurd extreme that every perspective is regarded as good as any other. In the science wars, scientific objectivists liked to portray their enemies as holding such a view, thus making *perspective* a dirty word. I therefore need to make it clear at the start that a *scientific* perspectivism does not degenerate into a silly relativism.

A more auspicious point of departure is the idea of viewing objects or scenes from different places, thus producing different visual perspectives on said objects or scenes. Visual perspectives possess an intersubjective objectivity in that there is roughly a way something looks from a particular location for most normal viewers. Any normally sighted person who has toured Washington, D.C., for example, can easily imagine how differently the Washington Monument looks from the Capital Building and when looking up at it from near its base.

Here there are resources for a naturalistic account of perspective beginning with the Renaissance theory and practice of linear, or "one point," perspective as rediscovered early in the fifteenth century by Brunelleschi and first codified by Alberti in 1435. Several later twentieth-century studies of the Renaissance use of perspective have already found their way into the contemporary science studies literature (Ivins 1973; Alpers 1983; van Fraassen 2004). In spite of its many tempting features, I will not pursue this topic here. I will, however, exploit findings about human vision, where the notion of perspective surely originated.

My prototype for a scientific perspectivism will be *color vision*. Although this is not a common way of speaking in English, it is perfectly clear what one would mean by saying that the human perspective on the world is typically colored. We typically see objects in the world as being colored. Moreover, like spatial perspectives, color perspectives are intersubjectively objective. That is, most people generally see the same objects as similarly colored in similar circumstances. Whether colors are objective in the stronger, more technical, sense of objectivist realism remains to be seen. I will argue that they are not. Colors are real enough, but, I will be claiming, their reality is perspectival. And it is perspectival realism that provides us with a genuine alternative to both objectivist realism and social constructivism.

I will assume that the considerations suggesting that color vision is perspectival can be extended to human perception more generally. What is perhaps more controversial is the extension to scientific observation. Most observational data in the sciences is now produced by instrumentation, sometimes very complex instrumentation. I will try to show that the output of instruments is perspectival in much the way that color vision is perspectival. Here we can distinguish two dimensions to the perspectival nature of claims about the output of instruments. First, like the human visual system, instruments are sensitive only to a particular kind of input. They are, so to speak, blind to everything else. Second, no instrument is perfectly transparent. That is, the output is a function of both the input and the internal constitution of the instrument. Careful calibration can reduce but never eliminate the contribution of the instrument.

More controversial still is the extension of perspectivism to scientific theorizing. I will try to show that the grand principles objectivists cite as universal laws of nature are better understood as defining highly generalized models that characterize a theoretical perspective. Thus, Newton's laws characterize the classical mechanical perspective; Maxwell's laws characterize the classical electromagnetic perspective; the Schrödinger Equation characterizes a quantum mechanical perspective; the principles of natural selection characterize an evolutionary perspective, and so on. On this account, general principles by themselves make no claims about the world, but more specific *models* constructed in accordance with the principles can be used to make claims about specific aspects of the world. And these claims can be tested against various instrumental perspectives. Nevertheless, all theoretical claims remain perspectival in that they apply only to aspects of the world and then, in part *because* they apply only to some aspects of the world, never with complete precision. The result will be an account of science that brings observation and theory, perception and conception, closer together than they have seemed in objectivist accounts.

Perspectivism makes room for constructivist influences in any scientific investigation. The extent of such influences can be judged only on a case-by-case basis, and then far more easily in retrospect than during the ongoing process of research. But full objectivist realism ("absolute objectivism") remains out of reach, even as an ideal. The inescapable, even if banal, fact is that scientific instruments and theories are human creations. We simply cannot transcend our human perspective, however much some may aspire to a God's-eye view of the universe. Of course, no one denies that doing science is a human activity. What needs to be shown in detail is *how* the actual practice of science limits the claims scientists can legitimately make about the universe.[31]

Finally, reflexivity poses no special problems for perspectivism. Consensus among scientists on a particular scientific perspective arises out of *both* social interactions among members of a scientific community and interactions with the world, typically mediated by complex instrumentation. But just as scientists do in this way succeed in creating more detailed or more general, or otherwise more desirable, perspectives on the world, so those of us who study science as a human activity can do the same. It is the best any of us can do.

Retrospect

Influenced by refugees from German-speaking Europe, the philosophy of science in the United States from roughly 1940 to 1960 focused on the *products* of scientific activities, particularly scientific theories, and on general methodological issues such as the nature of confirmation and explanation. The general method of the philosophy of science itself was one of conceptual analysis, often accompanied by reconstructions of both theory and method in the idiom of formal, symbolic logic. The history of science was largely intellectual history, again focusing on the development of scientific theories. The sociology of science focused on the structure and functioning of the social system of science.[32] In retrospect, and in spite of several apparent counterexamples,[33] the general attitude toward scientific claims was one of objectivist realism.

Beginning in the 1960s, inspired by Thomas Kuhn's *Structure of Scientific Revolutions* (1962) and other works (Hanson 1958; Toulmin 1972), attention gradually shifted to the *process* of doing science. Some philosophers began looking to the history of science (Lakatos 1970; Laudan 1977), historians looked more and more to social history, and sociologists looked to the content of the sciences relative to their social context.[34] The result has been that, by the beginning of the twenty-first century, most scientists and many philosophers of science retain an objectivist view of science, while many historians and most sociologists of science presume some more or less explicit form of constructivism.

The strongest support for constructivism has been in the form of a myriad of case studies, both historical and contemporary, though often a constructivist approach is assumed rather than argued. Scientists and their philosophical supporters are largely unmoved by these case studies, whether historical or contemporary. They feel they know that objectivism is correct and that the case studies cannot possibly show that constructivism is right.

Given this view of things, my goal is to provide a general argument, based on known features of scientific practice and of human cognition, that some degree of contingency is always present in science. So constructivists are at least partly right. If scientific knowledge is perspectival, scientific claims are neither as objective as objectivist realists think nor as socially determined as even moderate constructivists often claim. In fact, my main arguments are directed at showing that perspectival realism is as much realism as science can provide. Objectivist realism cannot be even an ideal goal.[35]

I realize that many people on the scientific side of the science wars regard the rejection of objectivist realism and the possibility of any degree of social constructivism as undercutting the reforming power of science in human affairs.[36] My view is just the opposite. By claiming too much authority for science, objective realists misrepresent science as a rival source of absolute truths, thus inviting the charge that science is just another religion, another faith. A proper understanding of the nature of scientific investigation supports the rejection of all claims to absolute truths. The proper stance, I maintain, is a methodological naturalism that supports scientific investigation as indeed the best means humans have devised for understanding both the natural world and themselves as part of that world. That, I think, is a more secure ground on which to combat all pretenses to absolute knowledge, including those based on religion, political theory, or, in some cases, science itself.

CHAPTER TWO
COLOR VISION

Introduction

Color vision provides the best exemplar I know for the kind of perspectivism that characterizes modern science. It is a phenomenon with which almost everyone is familiar. Moreover, the basic science of color vision is quite accessible. So one can here easily follow a naturalistic methodological stance. Of course, the phenomenon of color vision has been a staple of empiricist philosophers since Locke. It has also recently become a topic of considerable interest among contemporary analytic philosophers.[1] Nevertheless, I prefer to begin with the contemporary *science* of color vision.

Basic Color Science

Most humans experience the world as apparently containing colored objects as well as other colorful phenomena such as sunsets. The range of human color vision is familiar from the spectrum produced by rainbows. In the spectrum, the colors range from a deep purple to a dark red. It has long been known that these colors are in some way related to electromagnetic radiation with different wavelengths, ranging from roughly 400 to 700 nanometers (1 nm = 10^{-9} meters, that is, one-billionth of a meter). This is but a tiny fraction of the total electromagnetic spectrum, which includes cosmic rays with wavelengths around 10^{-14} meters and radio waves with wavelengths around 10^{6} meters. See plate 1 in the color insert.

In the nineteenth century, the German physiologist Ewald Hering showed that one aspect of our perceptual color space has a quite definite structure exhibited in what is now called a hue circle (plate 2 in the color insert). A continuous version of this circle contains all the *hues* perceived by humans. In addition, colors differ in *saturation*, or intensity, and *brightness*, that is, relative lightness or darkness. These three aspects of color can be represented in a three-dimensional array, such as the Munsell Solid, which represents all the colors that can be perceived by humans. For most of my purposes here, it will be sufficient to concentrate on hue.

There are four *unitary* hues: red, green, blue, and yellow. Thus, all the hues in the upper half of the hue circle are somewhat reddish and all those in the lower half are somewhat greenish. Similarly, all those to the left are somewhat bluish, and all those to the right are somewhat yellowish. In between any two unitary hues are *binary* combinations of these two unitary colors, such as the orange hue between red and yellow. From the structure of the hue circle, we can see that there are no such colors as a reddish green or a greenish red or, likewise, a yellowish blue or a bluish yellow. Red and green, and likewise blue and yellow, exclude each other.

The fact that hues have a circular rather than linear structure means that there is no simple linear relationship between wavelength and color. Hues ranging from purples through magenta to deep red are called "nonspectral hues" because they do not correspond to any single monochromatic wavelength. Rather, these hues can be produced only by a combination of two or more different wavelengths of light (Palmer 1999, 98). Thus, no simple one-to-one identification of perceived hues with single wavelengths is possible.

Opponent-Process Theory

Humans are *trichromats*, which is to say, their retinas contain three different types of receptors (called cones for their shape when viewed through a microscope) with three different pigments sensitive to three different ranges of the visible spectrum.[2] These three pigments, conventionally labeled S(hort), M(edium), and L(ong), have their peak sensitivities at roughly 440 nm, 530 nm, and 560 nm, respectively, although there is considerable overlap. These are shown in figure 2.1, arbitrarily scaled to unit sensitivity.

These three types of cones are neither equally numerous nor equally distributed in the retina. The ratio of S- to M- to L-type cones is roughly 1:5:10, with M and L types dominating in the center of the visual field. For any small region of the retina, the intensity of the signal from the various cones is transmitted to opponent cells both in the retina and the brain (particularly the lateral geniculate nucleus) that transform these signals according to the schema shown in figure 2.2, where *L, M,* and *S* now represent the total activation levels of the respective cones. Thus, what gets transmitted to the visual cortex for color vision is *differences* in the activations of the three

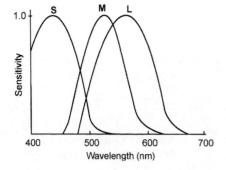

Figure 2.1 Relative spectral sensitivity of color pigments for trichromats.

types of pigments organized into two opponent systems, a red-green system and a yellow-blue system, which together produce the experience of all color hues.[3]

The nature of this transformation provides an immediate explanation of the overall structure of the hue circle. When $(L - M) > 0$, one experiences colors in the top half of the circle. Whether one's color experience is then in the upper-left or upper-right quadrant depends on whether $[(L + M) - S]$ is greater or less than zero. Similarly, when $(L - M) < 0$. The whole explanation is similar if one begins with $[(L + M) - S]$ being greater or less than zero.[4]

Opponent-process theory also partially explains the long-known phenomenon called *color constancy*, which refers to the fact that, for humans, the colors of things appear pretty much the same at widely variant levels of illumination, from low-level indoor lighting to bright sunlight. This is just what is to be expected if it is *differences* in the stimulation of different chromatic detectors rather than *absolute levels* of stimulation that drive the chromatic visual system.[5]

The Chromatic-Response Function

Because the human chromatic system is an opponent system, the relative response of an actual person's visual system to light all along the visible spectrum can be measured using a cancellation technique. The subject looks at a neutral surface through an eyepiece in an optical system that can mix colors

$(L - M) > 0 \implies$ RED

$(L - M) < 0 \implies$ GREEN

$[(L + M) - S] > 0 \implies$ YELLOW

$[(L + M) - S] < 0 \implies$ BLUE

Figure 2.2 Neural code for processing pigment activation in trichromats.

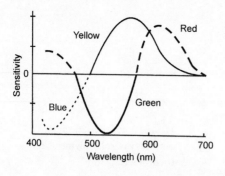

Figure 2.3 Chromatic-response function for trichromats.

within narrow bands of wavelengths. Consider just the yellow-blue system. One first determines the frequency of light for which the person reports seeing a pure blue, that is, a blue that exhibits no mixture of either red or green. This will be around 475 nm. Then one projects a light around the shortest wavelength reported to exhibit any yellow hue. At this point, one adds the pure blue light of fixed intensity and the subject adjusts an absorbing filter on the fixed blue light until the yellow disappears. Knowing the intensity of the two lights and the calibration on the filter, it can be determined how intense the pure blue light needs to be to cancel out the yellow response. This procedure is repeated for increasing wavelengths, say at 10 nm intervals, up to the end of the visible spectrum around 700 nm. At first the intensity of blue light needed to cancel out the yellow will increase to around 580 nm and will then drop off toward zero for longer wavelengths. The result is a curve giving the relative sensitivity of the person's visual system to the full range of wavelengths perceived as to any degree yellow.

This general procedure is repeated, this time canceling blue hues using a pure yellow light, and then repeated two more times for red and green sensitivities, respectively. The result is four curves showing the relative sensitivity of the subject to each of the four unitary hues as a function of wavelength. To represent the two opponent systems, the convention is to count red and yellow sensitivity as positive and then to assign negative sensitivities to green and blue. All four curves are then combined into one overall *chromatic-response function*, as shown in figure 2.3. These curves show the relative response to the four unitary hues as a function of wavelength.[6]

From these curves, we can immediately recover the pure primary colors, green, yellow, and blue. Moving up from shorter to longer wavelengths, the red-green curve shows a null response at roughly 475 nm, which would result in the experience of a pure blue. Next, the yellow-blue curve reaches zero at roughly 500 nm, so the response is a pure green. Finally, the red-green curve again crosses zero at roughly 580 nm, which results in a pure

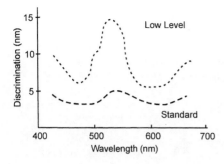

Figure 2.4 Wavelength discrimination as a function of wavelength and intensity.

yellow. Since the yellow-blue curve only approaches but does not cross zero at higher wavelengths, there is, contrary to common belief, no single wavelength corresponding to a pure red. All single wavelength reds have some blue or yellow added.

The chromatic-response function shown in figure 2.3 is idealized with respect to the accuracy with which humans can discriminate differences in wavelengths. As shown in figure 2.4, for normal subjects the ability to discriminate wavelengths varies both as a function of wavelength and the intensity of the light. For normal illumination, discrimination varies from plus or minus 3 nm at around 500 nm and 600 nm to plus or minus 5 nm around 550 nm and also at the limits of the visible spectrum around 400 nm and 700 nm. At low light levels, discrimination is much poorer, varying from plus or minus 6 nm around 450 nm and 600 nm, to plus or minus 15 nm around 530 nm, and plus or minus 10 nm at the limits. These levels of discrimination are an inherent feature of the human chromatic visual system.

Metamerism

Opponent processing and the chromatic-response function of normal viewers provide a basis for understanding the well-known phenomenon of metamerism, that is, the production of the *same* color experience by light with *different* spectral characteristics. For example, monochromatic light with wavelength around 580 nm projected on a neutral screen will be experienced as a pure yellow (solid vertical line in figure 2.5). The same color experience, however, can be produced by an appropriate intensity mixture of two monochromatic lights with wavelengths 540 nm (a greenish yellow) and 640 (a reddish yellow), represented by dotted lines in figure 2.5. Here, the green and red signals cancel one another, leaving the pure yellow (Hurvich 1981, 48). And this is just one out of a potential continuum of pairs of monochromatic light yielding the same experience of pure yellow. Similar examples exist for pure blue (475 nm) and pure green (500 nm).

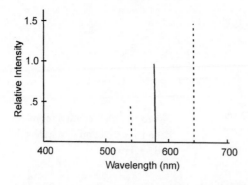

Figure 2.5 Metamers for pure yellow.

What are called *complementary* wavelengths are a special case of metamerism in which the mixing of two projected monochromatic lights produces a colorless light on a neutral screen. For example, a yellowish red at 670 nm and a bluish green at 490 nm yield a colorless overlap. Similarly, the overlap of a yellow at 580 nm with a blue at 480 nm is again colorless (Hurvich 1981, chap. 8). Many other pairs of monochromatic light are complementary in this sense.

Finally, all possible perceived color hues can be produced in a normal viewer by combining just three monochromatic sources at appropriate intensities, for example, 440 nm (a reddish blue), 540 nm (a yellowish green), and 650 nm (a yellowish red). Many other combinations of three sources can accomplish the same feat (Hurvich 1981, 94–95; Palmer 1999, 101–5). This phenomenon is exploited to produce color television.

Color Naming

Although they are fast dying out, there have been many studies of cultures that do not have words corresponding to many of the color terms found in modern European languages. Indeed, there are records of a few societies that have only *two* color terms, which can be characterized in English as roughly light-warm (including white, red, and yellow) and dark-cool (including black, green, and blue). Until the last third of the twentieth century, this diversity in color naming was taken as a prime example of cultural relativism, as championed by the linguists Edward Sapir and Benjamin Lee Whorf. As late as the 1960s, one can find linguistic textbooks expressing the Sapir-Whorf hypothesis that which parts of the spectrum get names is culturally contingent:

> Consider a rainbow or a spectrum from a prism. There is a continuous gradation of color from one end of the spectrum to the other. . . . Yet an American describing it will list the hues such as *red, orange, yellow, green,*

blue, purple, or something of the kind. The continuous gradation of color which exists in nature is represented in language by a series of discrete categories. This is an instance of structuring of content. There is nothing inherent either in the spectrum or in the human perception of it which would compel its division in this way. The specific method of division is part of the structure of English. (Gleason 1961, 4)

In the 1960s, Brent Berlin and Paul Kay (Berlin and Kay 1969) studied color naming among members of twenty different cultures. Using independent linguistic criteria, they identified roughly only a dozen so-called basic color terms. They found that even individuals in cultures with only two, three, or four color terms could be taught to discriminate among basic colors. Moreover, they found a systematic progression as languages exhibited more basic color terms. When one gets to languages with six basic color terms, they included red, yellow, green, and blue, plus white and black. These are the four unitary colors of the hue circle and opponent-processing theory. This, they reasoned, cannot be an accident of cultural history. Thus, it seems that the perception of color bands (red, orange, yellow, green, blue, purple) across a continuous spectrum is not merely a cultural and linguistic construct, but also partly a function of the human visual system.

Color Subjectivism

One of the most quoted passages in the literature on color vision is Galileo's expression in *The Assayer* of what I will call color subjectivism. "I think that tastes, odors, colors, and so on are no more than mere names so far as the object in which we place them is concerned, and that they reside only in the consciousness. Hence if the living creatures were removed, all these qualities would be wiped away and annihilated" (Drake 1957, 274). Similar views can be found in the writings of other major figures in the Scientific Revolution, including Descartes and Newton, and later, of course, Locke (Thompson 1995, chap. 1).

A similar view can be found in contemporary textbooks on color vision. Here is but one example:[7] "People universally believe that objects *look* colored because they *are* colored, just as we experience them. The sky looks blue because it *is* blue, grass looks green because it *is* green, and blood looks red because it *is* red. As surprising as it may seem, these beliefs are fundamentally mistaken. Neither objects nor lights are actually 'colored' in anything like the way we experience them. Rather, color is a *psychological* property of our visual experiences when we look at objects and lights, not a *physical* property of those objects or lights" (Palmer 1999, 95). Here, the contrast between psychological and physical properties is stated about as clearly as possible.

Color subjectivism is nicely illustrated by the rare phenomenon of *synesthesia*, in which perceived shapes, sounds, or tastes elicit color experiences. Sometimes these experiences are quite stable, so that, for a particular person, looking at a large letter *M* will regularly elicit a yellow response. For someone else, the letter *M* might always appear blue. The days and months on a calendar might regularly elicit particular colors, such as, yellow for Wednesday, turquoise for Thursday. In these sorts of cases, it is clear that the experience of color is internally generated. Sounds are not colored, but they may produce visual experiences of color in people with an unusual sensory system.

Even for synesthetes, however, we regularly distinguish between color *appearance* and color *reality*. Although the stage during an orchestra concert may appear to a synesthete to be illuminated with undulating colored lights, we think we know it is really illuminated with regular white lighting. Turning to a more ordinary example, we think we know someone's red BMW is really red even though it does not appear red at night in a parking lot illuminated by sodium vapor lamps. Color seems no different from other physical properties, like shape. The BMW might look like a Toyota from a certain angle, but it is still really a BMW.

One might go on to argue that Galileo and his modern followers are victims of a simple confusion. If sentient beings had never evolved on Earth, there would, of course, be no *experiences* of seeing a blue sky. But surely the color of the sky does not depend on the accidents of evolutionary history. Color subjectivism seems easily refuted by simple common sense. But common sense can be mistaken, or at least seriously misleading.

Common sense tells us that the sun rises in the east and sets in the west. That is how we talk about it in many languages. But since the Scientific Revolution we know that it is an illusion that the sun rises. Rather, the Earth is turning. We just don't normally have any independent experience of it turning. The ground beneath our feet seems fixed. Galileo could well have thought that our perception that objects have colors is as much an illusion as our perception that the sun rises. Common sense tells us how the world appears. Science tells us something else.[8] Nevertheless, our commonsense way of talking rarely gets us into trouble. The scientific view of the world makes little difference to our everyday activities.

This line of argument can be extended. We experience objects as colored. But this is at least partly explained by the fact that the wavelength of visible light is very short relative to the size of ordinary objects. Therefore, the light reflected from objects stays together in bundles that retain their direction from the source to the observer. Consider sound waves. Many computers now use three-way speaker systems. The two mid- to high-frequency speakers provide directional information, so the sound coming from the right can be distinguished from that coming from the left. Not so with the one low-frequency speaker. It can be almost anywhere in the room, even under the desk. The low-frequency sound waves just fill the room, giving little indication the location of the source. If visible light had the wavelengths of low-frequency sound, we would not be so inclined to assign colors to objects.

A similar argument may be made with respect to color constancy. Because the colors associated with objects appear relatively constant under wide variations in the light intensity, they seem really to belong to the objects. Indoors or outside in bright sunlight, my green sweater maintains its green appearance. Nevertheless, it does not follow that its color is an intrinsic property of the sweater. The upshot is that, while color subjectivism conflicts with widespread and strongly held commonsense beliefs, this conflict need not be taken as providing a decisive refutation. There are good scientific explanations for why commonsense beliefs should be what they are even though they conflict with a scientific account of color vision.

Color Objectivism

One obvious way to develop a commonsense account of color vision is to try to treat colors as objective properties of things in the world; objects, of course, but also things like the sky, sunsets, and rainbows. This turns out not to be as simple as it might first seem. From a scientific point of view, the objective color of an object could not be an autonomous property of its surface. In principle, a specification of the molecular structure of a surface provides all the information needed to determine how that surface would reflect light of any spectral composition. In other words, the nature of the light reflected off an object is completely determined by the spectral composition of the incident light and the molecular makeup of the surface (plus, perhaps, the angle of incidence). It cannot be that, in addition, there is another physical property that is the color.[9] So an objectivist regarding colors is bound also to be a physical reductionist regarding colors.

Nevertheless, one would not want to try to identify colors with these basic physical properties, and for several good reasons. One is that many physically very different surfaces would have to be assigned the same objective color. For example, a rose and a good color photograph of the rose should be pretty much the same color. Yet the surface of the rose and that of the photograph are physically very different. It seems highly unlikely that

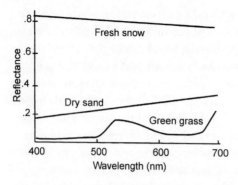

Figure 2.6 Surface spectral reflectances for several common surfaces.

fundamental material science contains any concepts that would classify these two surfaces as being of the same kind in any meaningful physical respect.

A second reason for not identifying colors with the material properties of surfaces is that the *perceived* color of a surface depends only on the spectral makeup of the light that leaves that surface and not at all on the deeper physical nature of that surface. There need only be a systematic way of characterizing how a surface reflects light, and that is provided by what is called its *surface spectral reflectance*.

Imagine shining on a surface a light of constant intensity all across the visible spectrum. Then, for every wavelength (or small interval), one measures the fraction of the incident light reflected at that wavelength. So the spectral reflectance of a surface at any particular wavelength varies between zero and one. The curve representing spectral reflectance all across the visible spectrum is the surface spectral reflectance of that surface. Figure 2.6 shows the surface spectral reflectance for several common surfaces. So now we have at least one clear proposal for an objectivist account of color, advocated, for example, by Hilbert (1987). The color of an object is its surface spectral reflectance, which is defined in purely optical terms.

A major problem with this account is the existence of metamers for spectral reflectances. Just as the perceived color of any monochromatic light source can be duplicated by mixing two different monochromatic lights, the perceived color associated with any surface spectral reflectance can be duplicated by innumerable other surface spectral reflectances. Figure 2.7 shows the relative intensities as a function of wavelength for three very different surface spectral reflectances that yield the same yellow appearance. What do these surface spectral reflectances have in common? Only that they are perceived as having the same color by normal trichromats. There is no category within optical science itself that would theoretically group together these spectral reflectances.

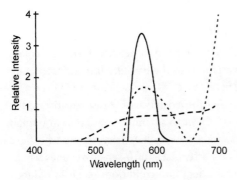

Figure 2.7 Metameric surface spectral reflectances.

So here is the situation. For any surface spectral reflectance, we can determine the approximate perceived color by a normal viewer by matching up the surface spectral reflectance with the chromatic-response function for normal viewers. The question is: How will this surface spectral reflectance interact with the known chromatic-response function to produce a color response? And this question always has an answer. Does this mean that colors are objective properties of the world? Yes and no.

If one considers the objective world to include humans as part of nature, then colors are an objective part of the world. This is in line with a commitment to naturalism. But if one imagines a world in which humans never appeared with their contingently evolved visual system, then there is no basis for assigning colors to any object. Why should a surface with a given surface spectral reflectance be called "yellow"? Without reference to the particular characteristics of the human visual system, there is no physical basis whatsoever for this identification.[10]

I have here examined only one form of objectivism, but I think any form of objectivism will run into the same problems. One can find physical properties such that, if there were humans around, they would have predictable color experiences. But there are many different physical properties that would give the same color experience. Thus, without reference to the specific nature of the human visual system, one has just a hodgepodge of physical characteristics.[11]

Comparative Color Vision

It helps to understand color vision, I think, if it is considered in its comparative and evolutionary contexts. Color vision evolved, and it evolved in different forms, in different species, at different times, and in different contexts. I will begin with comparisons within the human species and then move on to other species.[12]

Variations in Human Color Vision

There is a range of genetically based color vision deficits (forms of "color-blindness") that correspond to variations in the basic *trichromatic* structure of the human visual system. The prevalence of cones sensitive to Short wavelengths is genetically quite stable. But the existence of cones sensitive to the Medium and Long wavelengths is subject to sex-linked variation in human males. A mild form of deficit occurs in anomalous trichromats who have either two sets of M-sensitive cones or two sets of L-sensitive cones whose peak sensitivities are quite close together. Both conditions result in reduced sensitivity to reds and greens. More serious is loss of either the M- or the L-sensitive cones altogether. Figure 2.8 shows the sensitivities of the S-sensitive and L-sensitive cones for a dichromat. The result is *dichromatic* color vision, commonly known as total red-green colorblindness, in which one experiences only somewhat faded yellows and blues. It is estimated that about 8 percent of Caucasian males exhibit one or another of these various forms of colorblindness. Plate 3 (in the color insert) shows a simulated comparison of a scene as seen by a normal human trichromat and by a human dichromat. The chromatic aspects of the scene are much diminished for the dichromat.

Far less common is *monochromacy*, a condition in which there is only one set of color-sensitive cones and thus no possibility of any opponent-processing. The result is no color experience at all, but maybe some ability to distinguish chromatic features of light due to interaction among cones and rods. Finally, there is total *achromatopsia*, which is due to a failure of the neural processing system for color signals, resulting in no chromatic sensitivity whatsoever. Visual experience ranges from black through all shades of gray to white.[13]

The same genetic mechanisms that produce red-green colorblindness in men allow the possibility that some mothers of colorblind men might be *tetrachromats* with an additional set of cones sensitive to wavelengths

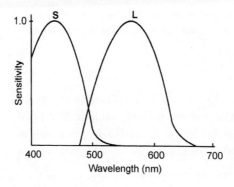

Figure 2.8 Relative spectral sensitivity of color pigments for dichromats.

around the wavelengths for standard Medium and Long light-sensitive cones. Such a woman would be able to perceive color differences in objects that appear identical in color to normal trichromats, in much the same way that trichromats can perceive color differences not accessible to dichromats (Thompson 1995, 167–68).

The Evolution of Color Vision

The oldest and most selective color vision systems are not those of humans, primates, or even mammals, but those of fish and birds. Among fish, such as the common goldfish, and birds, such as ducks, tetrachromacy is common, with the fourth pigment having peak sensitivity in the near ultraviolet. There is even some evidence that some species, such as the pigeon, are pentachromats. Also, the *range* of chromatic sensitivity is typically greater in these species than in primates, as broad as from 350 to 800 nm in salmon, that is, a range of 450 nm as compared with 300 nm for humans. Finally, in some of these species, cones come equipped with colored oil drops that act as light filters, thus sharpening the range of wavelengths over which the associated cones are sensitive. No mammals have been found with this particular adaptation.[14]

Among the few groups of species that have been studied in some detail, one finds the expected adaptation of their chromatic visual systems to their respective environments. Fish that live at greater depths, for example, show peak sensitivity to the wavelengths that predominate at their respective depths. Hummingbirds are especially sensitive to the ultraviolet light strongly reflected by the flowers on which they feed. Similarly, the ultraviolet sensitivity of some birds corresponds to the fact that plants whose berries they eat reflect strongly in the ultraviolet region of the spectrum (Jacobs 1993, 455).

Among mammals, by contrast, most species that have been shown definitely to have color vision are only dichromats.[15] This is true, for example, of the much-studied domestic dog and cat. Only humans, and their ancestors among the Old World monkeys and apes, exhibit robust tetrachromacy. If mammals evolved ultimately from fishes, why should the chromatic visual system of mammals be so impoverished relative to that of fishes? The answer is that the earliest mammals were nocturnal, thus favoring a visual system rich in rods rather than cones. So chromatic sensitivity declined. Up until the 1980s, it was widely believed that most mammals had no color vision at all. More recent research shows that cones did not disappear in mammals, although their density did greatly decrease compared with that of rods. So color vision did not have to reevolve for mammals. It only needed the proper selective environment. The currently accepted wisdom is that trichromatic color vision evolved among diurnal Old World monkeys in an environment in which it was advantageous to be able to distinguish ripe red fruits from green background vegetation (Mollon 2000).

Problems for Color Subjectivism and Color Objectivism

A comparative evolutionary understanding of color vision creates problems for both color subjectivism and color objectivism. The problem for color subjectivism is that the idea of fitness, or selective advantage, makes sense only as a *relationship* between organisms and their environments. An evolutionary biologist would want to say, for example, that monkeys who can more reliably *distinguish* ripe orange fruit from unripe green fruit have a selective advantage favoring emerging trichromats over dichromats. A color subjectivist, however, would have to say that the advantage goes to those monkeys who can more reliably *correlate* their differing color sensations with ripe and unripe fruit respectively, and trichromats are better equipped than dichromats to establish the requisite correlation. While not absurd, this latter seems a scientifically implausible characterization of what the monkeys are in fact doing. The scientifically accepted way of describing the situation is to say that the more advanced chromatic visual system enables its possessors better to detect relevant differences in the environment. It is not a matter of correlation but of causation.

The problem for color objectivism is a version of that discussed earlier. Whatever the color objectivist identifies as real colors are unlikely to be what the evolutionary biologist would identify as the relevant environmental selective factor. Consider again a color objectivist who identifies colors with surface spectral reflectances. This objectivist would have to say that the color visual system evolved better to detect spectral reflectances. The evolutionary biologist would say that the visual system evolved better to detect ripe fruits. And the fact is that different ripe fruits that we would identify as of a similar orange color will have differing spectral reflectances. Thus, even though the visual system is responding to spectral reflectances, the relevant set of reflectances is determined not by any distinctive physical characteristics of the reflectances themselves, but the nutritional value of the objects producing those reflectances. That is to say, the selective factor is causally identified with the similarity in perceived color of the ripe fruits, not the spectral reflectances. A similar argument would apply to any other objectivist definition of colors.

There is a further problem with the particular identification of colors with surface spectral reflectances. For many fishes, the color of surfaces may be less important than the *contrast* between the relatively uniform blue of the water and the image of likely prey or predators. So the selective factor in the environment is not reflectances at all. The same may be true of some birds that feed on flying insects set against the background of foliage or of the sky. In these cases, it is color contrasts, not reflectances, that are doing the selective work in the evolution of the respective visual systems.

A Possible Lesson from the Comparative Study of Color Vision

A comparative study of color vision leads naturally to consideration of the evolutionary development of different chromatic visual systems. But an evolutionary analysis cannot focus just on the internal workings of an organism or just on the external environment. It is the *interaction* between organisms and their environment that drives evolution. Perhaps the best way to understand color vision is to focus less on either the physical world or the organism and more on the interaction between the two.

Thinking about the interaction between the physical world and the organism suggests what might be the underlying problem with both color subjectivism and color objectivism. Both seek to understand colors as properties of one kind of thing. For objectivists it is some feature of light or surfaces. For subjectivists it is a feature of the visual system. A simple, almost obvious, suggestion is that it is a feature of both. In philosophical terms, perhaps color terms are best understood not as *monadic* but as *relational* predicates (Thompson 1995).[16]

Color Perspectivism

A very good statement of what seems to me the right thing to say about color vision appears already in Hurvich's pathbreaking text on color vision.

> It should be clear by now that object color is not physical light radiation itself, that it is not something that inheres in objects, having to do exclusively with the chemical makeup of the object, nor is it only the nervous excitation that occurs in the eye and brain of an observer. In our perception of object color all these elements are involved; there is light radiation, which is selectively absorbed and reflected in different ways by objects that differ physically and chemically; when the light rays coming from objects are imaged on the retina, they set off a complex series of neural events that are associated with the visual experience of color. (Hurvich 1981, 52)

The view expressed here could fairly be called "color interactionism." Colors are the product of an *interaction* between aspects of the environment and the evolved human visual system. *Interaction* is just the word used by Palmer. I earlier quoted his clear statement of color subjectivism. Immediately following the quoted passage, however, he continues: "The colors we see are based on physical properties of objects and lights that cause us to see them as colored, to be sure, but these physical properties are different in important ways from the colors we perceive. Color is more accurately understood as the result of complex *interactions* between physical light in the environment and our visual nervous system" (Palmer 1999, 95; emphasis added). My suspicion is that vision scientists tend not to distinguish between subjectivism and interactionism. Because the interaction may be regarded as partly subjective, that is enough to rule out straightforward objectivism. Assuming a traditional dichotomy between "objective" and "subjective," or, in Parker's earlier words, between "psychological" and "physical," they come down on the side of subjectivism.[17]

The view I prefer, *color perspectivism*, is an asymmetric version of color interactionism. I prefer the asymmetric version because I want to emphasize the human side of the interaction. I want to say that the typical human experiences the world from a colored perspective. We humans have a particularly human perspective on the world. The world has no particular perspective on us. It does not care about us.

Color perspectivism has the great advantage that it can incorporate the features of both color subjectivism and color objectivism that give those views their initial plausibility. It accommodates the subjectivism of synesthesia and drug-related chromatic experiences simply because the color visual system can be activated by activities internal to the brain itself. It also takes account of features both of the human visual system and of the world that make colors seem to be relatively permanent characteristics of many objects. But it leaves ample room for other, more transient, color experiences, such as viewing sunsets or rainbows. Even the familiar fact that the sky is blue can be well understood in perspectival terms.[18]

Needless to say, color perspectivism incorporates the findings of color science. And, like the science itself, it is open-ended regarding new findings. In particular, the neuroscience of color vision is currently a very active area of research. I have not reviewed this research here only because it does not seem necessary for my present purposes.[19]

Perspectivism also makes good sense of comparative studies of color vision. If humans have a particular colored perspective on the world, so do monkeys, birds, fish, cats, and dogs. The perspective is different for different species, but it is still a colored perspective, and we can learn about these different perspectives. For dichromats and other trichromats, we can even simulate what the world looks like to them, though, of course, we cannot simulate their subjective experience. For tetrachromats or pentachromats, we cannot even simulate their visual images. We have only the analogy of going from dichromatic to trichromatic color vision to supplement our descriptive knowledge of how their chromatic visual systems work.

The Compatibility of Visual Perspectives

I would now like to draw some general lessons about visual perspectives and, hopefully, about perspectives more generally. Let us compare the perspective of a typical human trichromat with that of someone suffering from complete red-green colorblindness and, thus, a person with dichromatic color vision. Suppose they are both viewing a rug that to the trichromat is green with a red pattern. The dichromat sees only a uniform faded blue. Can it be said that the visual experience of one is veridical and the other mistaken? I think not. There is no color that the rug is "really," that is, objectively. There is only the color of the rug as seen by a dichromat and the color as seen by a trichromat. If another trichromat claimed the pattern is more orange than red, then there might be a genuine disagreement. Similarly, if another like dichromat claimed the rug to be a uniform pale yellow. But the experience of the dichromat and the trichromat are not conflicting, just different.

The difference is, of course, a genuine difference. One way of expressing the difference is to say that the trichromatic visual system is capable of extracting not only different but *more* chromatic information from the environment. It follows that it is impossible to recover the chromatic experience of the trichromat from that of the dichromat. It does not follow, however, that it is possible to recover the dichromatic experience from the trichromatic. The existence of metamerically equivalent spectral reflectances shows that some information about the spectral reflectance reaching the eye of the trichromat is lost in the processing. One could not know how the unknown spectral reflectance would interact with a dichromatic system, therefore, even if one knew the chromatic characteristics of the dichromatic system itself. It would seem, however, that it is a general characteristic of chromatic visual perspectives that different perspectives are always *compatible*. There seems no way to generate a genuine inconsistency between different chromatic systems. They are just different.

Finally, a common objection to perspectivism of any form is that it may lead to an undesirable relativity. On a perspectival understanding of color vision, however, while there is relativity to a particular type of chromatic visual system, this relativity need not be objectionable. The trichromatic perspective is a widely shared, species-specific trait among humans, and, once acculturated into a linguistic community, individuals are highly constrained in their public color judgments. Thus, understanding objectivity as reliable intersubjective agreement, color judgments turn out to be quite objective.

The Uniqueness of the World as a Methodological Maxim

But can we not say more? Is not the reason why different visual systems do not produce genuine conflicts that they are all interacting with one and the same environment? In particular, is not the overall spectrum of electromagnetic radiation at a given time and place unique? Thus, given the same scene, different chromatic systems would produce different, but always compatible, images. The uniqueness of the world would guarantee the compatibility of different perspectives.[20]

Now, the uniqueness of the world is a clear example of what would typically be taken as a metaphysical doctrine. But it need not be so regarded. It can be understood as merely a methodological presumption. In the scientific investigation of the world, we presume there is a unique causal structure to the world. But we do not need to justify this presumption a priori. It does not function as a premise in our reasoning. It is a presumption of our actions, justified only after the fact if we succeed in reconciling apparently conflicting perspectives.

The history of the concept of determinism provides a useful parallel here. In the eighteenth and nineteenth centuries, determinism was often regarded as a metaphysical presupposition of the sciences. The acceptance of indeterminism in quantum theory in the early twentieth century was therefore regarded by many as a metaphysical crisis. It need not have been so regarded. Determinism could have been understood as merely a methodological prescription to look for deterministic models of natural phenomena. As such a prescription, it was very successful for a very long time and is still useful in many sciences. It failed for atomic and subatomic phenomena. This could have been regarded as at most a methodological crisis. More modestly, the acceptance of indeterminism at the atomic level need only have required a restriction of the original methodological maxim to more macroscopic phenomena.

It is, I admit, difficult to imagine circumstances in which scientists would feel obliged to limit the maxim of presuming a unique structure behind a given phenomenon. That would require something like concluding that different samples of the same radioactive isotope had different half-lives and that no further explanation of this difference was possible. Nevertheless, this failure of imagination does not justify elevating the maxim of presuming a single structure to the world to the status of a metaphysical doctrine. It need not be regarded as more than a well-entrenched maxim of scientific practice.

The Partiality of Perspectives

For my purposes, maybe the most important feature of perspectives is that they are always partial. When looking out at a scene, a typical human trichromat is visually affected by only a narrow range of all the electromagnetic radiation available. In particular, the nearby wavelengths in the ultraviolet and infrared are simply not visually detected.[21] And of course there is no possibility of visually detecting gamma rays or neutrinos.

In 1934 Jakob von Uexkull published a now scientifically dated article with the charming title "A Stroll through the Worlds of Animals and Men: A Picture Book of Invisible Worlds." In this work he develops a theory of what he calls an *Umwelt*, the aspects of the world that can be either perceived or acted upon by a particular organism.[22] He constructs the *Umwelt* of a number of organisms, ranging from ticks, to dogs, to humans. Uexkull's *Umwelt* is a more elaborate version of what I am calling a perspective. His is a dramatic presentation of the partiality of different perspectives. But there is a danger in his way of thinking that I want explicitly to avoid right from the start.

As even the title of his monograph indicates, Uexkull is tempted to think of various organisms as "living in different worlds." This could be a harmless metaphor, but it often is not.[23] Far better, I think, to reaffirm the methodological principle that there is only one world in which we all live. Given our differing biological natures, we naturally interact with different aspects of the world. In this sense, we view it from different perspectives. But we should regard them all as perspectives on a single world. We humans have the advantage that we have been able to build instruments that permit us greatly to enlarge and enrich the perspectives from which we can view, and interact with, the world.

There is one final pitfall to be avoided. As Hilary Putnam (1999), along with others, has argued, modern philosophy (meaning mostly epistemology) has long been trapped in the idea that we perceive our own representations of the world. When I say we have a colored perspective on the world, I do not mean that we experience colored representations of the world. I mean we experience an interaction with the world that, given our biological nature, results in our being able to distinguish objects and other phenomena by their apparent colors. To put it another way, we perceive aspects of the

world itself, which aspects being determined by our particular sensory capabilities. *How* this happens for colors is being explained by color science.

Note, finally, that, in experimenting with human subjects, color scientists
do not need to make any particularly philosophical assumptions about the
nature of minds or consciousness, or about relationships between conscious
experience and the world. They simply rely on a refined commonsense
understanding of human sensory capacities. Typical experiments require
subjects to match the apparent colors of patches of neutral surfaces in care-
fully controlled viewing situations. This is little different from the common experience of having one's eyes checked for corrective glasses. Here
the patient is seated, looking through a binocular-like apparatus at letters
projected on a neutral screen in a semidarkened room. The optometrist
switches back and forth between different lens combinations and asks: "Are
the letters sharper with 1? Or 2?" Perhaps repeating, "1? Or 2?" The patient
answers, perhaps with qualifications, saying things like: "1 is a lot sharper."
"2 is a little sharper." "I really can't tell any difference." And in most cases the
patient ends up with the appropriate lenses. Common sense, informed by
general knowledge of the typical patient's capabilities in the controlled situation of the optometrist's office, is sufficient.

The Philosophy of Color

Color and color vision have long been staples of modern philosophy,
Hume's missing shade of blue (1888, 6) being perhaps the most famous
example. In 1988 C. L. Hardin sparked a small renaissance in philosophical
thinking about color vision with his book, *Color for Philosophers: Unweaving
the Rainbow*. A notable feature of this book was the introduction of many
findings from the contemporary science of color vision, including opponent-
process theory. Hardin ended up arguing for a subjectivist account of color,
indeed, a subjectivism ultimately reducible to neurophysiology. Since then,
the bulk of the philosophical literature has been devoted to refuting subjectivism in favor of some variety of objectivism (Byrne and Hilbert 1997a).
A notable exception is Evan Thompson's (1995) *Color Vision: A Study in
Cognitive Science and the Philosophy of Perception*, which promotes an inter-
actionist account of color vision.

I will not enter further into these debates here. For my purpose, I have
already said all I think needs saying about both subjectivist and objectivist
approaches to color vision. That purpose, recall, is to show that, from within
a general scientific framework, scientific knowledge is perspectival in ways
strongly analogous to the way our knowledge of colors is perspectival.

Nevertheless, to allay some concerns of my philosophical readers, a few final general remarks are in order.

Primary and Secondary Qualities

Many contemporary discussions of colors take for granted the seventeenth-century distinction between primary and secondary qualities. In fact, this distinction, in its seventeenth-century context, is not unproblematic.[24] Indeed, it seems that the motivation for making the distinction may have been more to further the cause of the new mechanistic science than to advance philosophy, though the distinction between "science" and "philosophy" in this context is also problematic. The new mechanical philosophy of Galileo, Descartes, Boyle, and Newton sought to explain all things in terms of the size, shape, and motions of small corpuscles, properties that could be represented mathematically. This circumstance goes a long way toward explaining why these properties were designated as being "primary." All other properties, which are then explained by the "primary" qualities, are, by implication, only "secondary."

As Thompson (1995) also emphasizes, the early modern account of color tends to be ambiguous, even in the writings of a single figure, such as Locke. Sometimes colors (and secondary qualities in general) are portrayed as being purely subjective experiences. Other times they are identified with the "powers" (or "dispositions") of primary properties of the minute particles that make up macroscopic objects to produce these subjective experiences in humans. This difference might be interpreted as being parallel to the contemporary contrast between "subjective" and "objective" accounts of colors. The dispositional version might also be interpreted as subject to the additional ambiguity already noted between a generalized materialism (or physicalism) and a genuine physical reduction of colors to physical features of a world apart from any perceivers. It is in general scientifically correct that, in any specific context, the physical constitution of the light together with the physical operations of the human visual system determine the color experience of a normal viewer. But there is also no property recognized by modern physical science that can play the role of the hypothesized "powers." There is only a disjunction of physical properties whose only unifying characteristic is that they produce the same experience in normal viewers. Whether any passages in the classical writings can be understood as enunciating a relational, or perspectival, view of colors is beyond my competence to judge. I am confident only that it would be misleading (if not an outright mistake) to identify a relational view of colors with the claim that colors are merely "secondary" properties.

What Might a Modern Philosophy of Color Be?

As a philosopher of science, I would describe the philosophy of color as the philosophy of color science, thus placing it alongside the philosophy of physics or the philosophy of biology. But there are deep disagreements even among philosophers of science as to what philosophers of physics or biology should be doing. Some think that the philosophy of a science should be continuous with science in the sense that philosophers are attempting to solve some of the same (mostly theoretical rather than experimental) problems as scientists, the measurement problem in quantum theory or the nature of selection in evolutionary theory. I regard this approach as at best misguided, at worst arrogant. It is misguided in that it presumes that the recognition of problems and possible solutions is a purely abstract, theoretical matter that transcends the boundaries of disciplines with their own internal criteria for who gets taken seriously. Scientists will not take philosophers, or anyone else, seriously unless those persons first somehow establish themselves as members of the relevant scientific community, a very difficult, though not completely impossible, task. The approach is arrogant in that it presumes that someone who has not gone through the standard processes of education and acculturation into a scientific field can actually recognize and solve problems that scientists in the field cannot. What typically happens is that philosophers create a separate community, say philosophers of quantum theory, whose members communicate primarily with each other and interact only weakly with the community of theorists in physics. Many works on color vision written by philosophers exhibit these features. The analyses invoke methods derived from the philosophy of language, the philosophy of mind, metaphysics, and epistemology, methods foreign to color science.[25]

A more modest approach to the philosophy of color science is to describe the practice and results of color science in broader terms accessible to members of the science studies community (historians, philosophers, and sociologists of science) as well as educated laypersons. Here one takes the science at face value, but one need not necessarily accept the way scientists characterize or present their science to themselves or to the general public. Thus, for example, when color scientists claim that colors are "subjective," "psychological," or "processes in the brain," they are not presenting the science, but offering a more general characterization of their science. Color scientists do not own these more general characterizations and can be legitimately challenged by outsiders who may, in fact, better understand what is implied by these more general claims. It is in this spirit that I claim that colors are best thought of as neither completely objective nor purely subjective, neither as properties of either parts of the material world or of subjective experience, but as a property of an interaction between the material world and human observers.

My attitude toward a philosophy of color science applies also, of course, to the whole project of this book. I do not question the physics of physicists such as Weinberg or Glashow. I question only their general characterization of their enterprise in absolute objectivist terms as the discovery of true laws of nature. This characterization is not a part of physics but is an *interpretation* of their enterprise. It can legitimately be questioned by knowledgeable outsiders.

Truth and Commonsense Color Realism

Let us return, finally, to commonsense claims about colors. Some philosophers (Stroud 2000) are impressed by the fact that many commonsense claims involving colors, for example, "Ripe lemons are yellow," seem obviously *true*. From a commonsense perspective, anyone who would say such a claim is false obviously does not know much about lemons, or perhaps misunderstands the meaning of the term *lemon*. The prevalence of such claims in everyday life, Stroud argues, undercuts the conceptual possibility of what he calls an "unmasking explanation" of color vision, that is, of claiming that things are "not really colored." His argument is far too long and convoluted to summarize here.[26] Insofar as it relies on commonsense color judgments being true, however, it is questionable.

Most philosophers, including, I would surmise, Stroud himself, agree that the statement "It is true that ripe lemons are yellow" has exactly the same empirical content as the statement "Ripe lemons are yellow." The former is merely a metalinguistic way of making the same empirical claim as the latter. Rather than simply asserting that ripe lemons are yellow, one asserts the truth of the statement that ripe lemons are yellow. The ability to make this metalinguistic move is a useful, perhaps even necessary, resource for advanced languages. Whether it is anything more than that is a much debated philosophical issue (Künne 2003). But without making a whole lot more of the concept of truth than could be ascribed to common sense, Stroud's conclusion simply does not follow.

I have already pointed out features of human color vision (e.g., relatively short wavelengths for white light, color constancy, low incidence of color blindness among humans) that go a long way toward explaining why humans typically ascribe colors to objects without qualification. This practice is quite understandable even if we admit that such ascriptions are relative to a normal trichromatic human visual system. It takes a fairly deep scientific understanding of how the human visual system works to appreciate this relativity.[27] It is not part of common sense. But it is not, therefore, nonsense.

Many philosophy professors have seen how confused introductory students can become when asked the old question whether a tree falling in the forest makes a sound if there is no one near to hear it. This is not a question for which common sense is prepared. It hardly ever arises in everyday life. The standard answer is that the falling tree produces pressure waves that have physical effects, such as shaking pine cones off of nearby pine trees. Sounds, however, are produced by these pressure waves only if there is the right kind of perceiver in the vicinity, and the character of those sounds depends on the character of the auditory system in question, different for humans and wolves, for example. The situation, I am claiming, is similar for colors, although the commonsense identification of colors with objects is much stronger, for the sorts of reasons already given.

What holds for sounds and colors holds, I think, for common sense generally. Although its outlines are vague, and different in different cultures and at different times, there is something we can call a commonsense perspective on the world. It is a perspective from which most people start because of their upbringing before formal education. And it remains a perspective within which we mostly operate unthinkingly, even after we learn the advantages of other, especially scientific, perspectives. It does not follow, however, that the commonsense perspective creates impenetrable boundaries around what we can understand. We can transcend those boundaries, and have repeatedly since the seventeenth century.

A Final Question

It is time explicitly to raise a question that no doubt has already occurred to many readers. From what perspective do I make all the above claims about the science of color vision? The simple answer is: from the perspective of color science. This includes parts of materials science, optics, psychology, physiology, neuroscience, and more. The trouble with this simple answer is that it invites a further question. Even if it is granted that color vision is perspectival, that does not show that all the sciences involved in color science are perspectival. In particular, it does not provide any reason to reject an objectivist understanding of these sciences. And indeed it does not. That takes a separate argument. I propose, however, to postpone giving that argument until after I have extended the lessons of color vision to scientific observation generally.

CHAPTER THREE
SCIENTIFIC OBSERVING

Introduction

No one questions that the claims of contemporary scientists are in some sense based on the observation of nature. The problem is how to understand the scientific process of observing nature. The first step is to realize that virtually all scientific observation now involves *instrumentation*. In this chapter I will be arguing that observation using instruments is perspectival in roughly the same ways that normal human color vision is perspectival. Indeed, the human visual system can be thought of as an instrument for observing the world around us. The fact that this instrument was produced by organic evolution rather than conscious design is less important than its functional role in human interactions with nature.

Some historians have argued persuasively that the development of scientific instruments has been motivated in large measure by a desire to make scientific observation *objective* in the sense of reducing the role of individual subjective judgments in assertions of what has been observed.[1] I agree with this analysis. There are indeed many respects in which instruments are more reliable than human observers. Nevertheless, even though human color vision, for example, is more radically perspectival than most instrumental observation, instrumental observation remains perspectival in many ways.

In the most general sense, scientific instruments are perspectival in that they respond to only a limited range of aspects of their environment. Just as the human visual system responds only to electromagnetic radiation, so do ordinary microscopes or telescopes. These systems are all equally blind to cosmic rays and neutrinos. But even for those aspects of the world to which they do respond, the response is limited. The human visual system responds only to electromagnetic radiation in the visible spectrum. A camera responds only to that radiation to which its film, or, more recently, its digital sensors, are attuned. The same is true for microscopes or telescopes. Finally, even within their range of sensitivity, instruments, again like the human visual system, have some limitations on their ability to discriminate among inputs that are theoretically distinct. The relationship between inputs and outputs always remains to some extent a many-one relationship. The nature of this relationship is part of the perspective of any particular instrument.

In this chapter I will be considering instruments in two different sciences, one, astronomy, concerned with the very large and the second, neuroscience, concerned with something relatively small, the human brain. In both cases, I will indicate those respects in which the relevant instruments are perspectival. Something similar, I suggest, holds for scientific instruments generally.

Astronomy in Color

In 1998 the National Academy of Sciences in Washington, D.C., hosted an exhibition titled "Night Skies: The Art of Deep Space." The exhibit consisted of forty color images of celestial objects, including stars, nebula, and galaxies—including even a picture of Halley's Comet. Plate 4 (in the color insert) shows one of these striking images, a picture of the Trifid Nebula. Here is part of the description that accompanied the image: "The spectacular Trifid nebula is one of the best known in the sky. It is a striking mixture of brilliant red light emitted from excited hydrogen gas and the soft blue glow of a reflection nebula. The blue arises from starlight, scattered by dust particles between the stars. The size of the particles is minute, similar to those of smoke, which also has a bluish hue. However, the scattered light is not a pure blue, and if we see it through a medium that is yellow (i.e., absorbs blue light) some green colouration remains." From this description it is clear that the color in this image is *true* color, that is, the Trifid here appears pretty much as it would appear to a human observer at an appropriate distance away. How was this accomplished? One does not see the color looking through an optical telescope. The light is not bright enough for that. Nor is color film sensitive enough faithfully to record such colors. Nor are current electronic detectors.

These pictures were made by David Malin, a photographer turned astronomer, at the Anglo-Australian Observatory in New South Wales, using a process originally invented by James Clerk Maxwell in the 1850s and perfected by Malin in the 1970s. He begins with highly sensitive ten-inch square black-and-white photographic plates. For each object to be photographed, he makes three separate exposures on three different plates. For one he uses a red filter, which filters out light in the green and blue parts of the visible spectrum, leaving only the red. This black-and-white negative is thus exposed only to the red part of the spectrum of light emitted by the source. Similarly, for the second exposure he uses a green filter, and for the third, a blue filter. Exposure times are around twenty-five minutes for each plate.

With each of these negative plates, he makes a positive contact print on black-and-white film. Then, using an enlarger, these positive black-and-white films are projected in sequence on a *single* sheet of *color* film. For the positive black-and-white film derived from the plate originally exposed with a red filter, this exposure is made using the same sort of red filter used to make the original plate. The color film thus records as red the back and white image originally formed by the red light from the source object. Similarly, the black-and-white film deriving from the original exposure using a green filter is projected using the same sort of green filter. Same for the blue. Finally, the resulting color negative is used to make positive color prints, like this one of the Trifid Nebula.

This whole process, I claim, is *multiply perspectival*. First, the original black-and-white plates are sensitive only to radiation in roughly the visible spectrum, that is, visible light. Each of the original three plates is then restricted to a still narrower perspective by its corresponding filter. Finally, all three perspectives are combined to produce a single image that we humans perceive as colored from the perspective determined by our capacity for color vision as described in chapter 2.

I was drawn to this example because it goes so well with my original example based on color vision in humans and other animals. But the point to be made is, I think, general. Scientific observation is always mediated by the nature of the instruments through which we interact with selected aspects of reality. In this sense, scientific observation is always perspectival. In this particular case, we do not simply have an observation of the Trifid Nebula "as it is in itself" so to speak. It is an observation of the Trifid *from the perspective provided by Malin's three-color process.*

Deep Space from the Perspective of the Hubble Telescope

Astronomers have dreamed of putting a telescope in space above the Earth's atmosphere since at least the 1920s.[2] The dream was realized on April 24, 1990, when the Hubble telescope was launched aboard the space shuttle Discovery. After a dramatic mission to correct an embarrassing flaw in its mirror at the end of 1993, the Hubble has produced genuinely revolutionary observations. In January 2003 the Space Telescope Science Institute (STSI) released a remarkable image produced by the Advanced Camera for Surveys (ACS) aboard the Hubble (see plate 5 in the color insert).

The process that produced this image is far too complex for me even to attempt to describe in any detail here. It involves electronic detectors sensitive to light in the infrared part of the electromagnetic spectrum. The output of the detectors is fed into an onboard computer and put into a form in which it can be transmitted to a tracking and data-relay satellite, from which it is retransmitted to the White Sands Complex near Las Cruces, New Mexico, from which it is again retransmitted by domestic satellite to the Data Operations Control Center at the Goddard Space Flight Center in Greenbelt, Maryland. From there it is routed to the Data Capture Facility and finally on to the Space Telescope Science Institute in Baltimore, where it is studied by astronomers and other space scientists. Each step in this process, sketched in figure 3.1, in some way modifies the initial signal and contributes to the construction of the image we see in plate 5.

It is worth noting that during the thirteen hours it took to accumulate the light for this image, four different filters were used at different times. Thus, this process employs an electronic version of the process invented by Maxwell and developed by David Malin. Four different components of the light, corresponding to four different ranges of wavelengths, are separated and then recombined to produce the final image.

Another remarkable feature of this particular image is that it involved gravitational lensing. During the exposure, the Hubble telescope was pointed directly at a massive cluster of galaxies known as Abell 1689, estimated to be 2.2 billion light-years away. In accordance with the General Theory of Relativity, this mass acts like a lens by warping space around it and thus effectively bending light passing by. Scientists who have studied the data claim that the image captures galaxies from which light was emitted

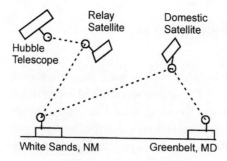

Figure 3.1 Path of the signal from deep space to the Space Telescope Science Institute.

roughly 13 billion years ago, when the universe was only one billion years old, which is to say, about one-fourteenth its present age.

Here again, I claim it is misleading to say simply that here we have an observation of the universe as it was 13 billion years ago. What we have is an image of the early universe *from the perspective of the Hubble ACS system using Abell 1689 as a gravitational lens*. The picture is the product of an *interaction* between light from the early universe and the Hubble telescope system.

The Milky Way in Gamma Ray Perspectives

NASA's second major satellite observatory, after Hubble, was the Compton Gamma Ray Observatory (CGRO), launched April 5, 1991, and deliberately deorbited June 4, 2000. Whereas Hubble's perspective on the universe is confined to electromagnetic radiation around the visible spectrum, the CGRO was designed to detect much more energetic gamma rays. The CGRO contained four different detectors which together covered the gamma ray spectrum from 30 keV to 30 GeV. The observatory was named for the American physicist Arthur Holly Compton because three of the observatory's four detectors used the phenomenon of Compton scattering (or the Compton effect), for which he received the Nobel prize in 1927. Thus, the physical processes on which the CGRO was based are fundamentally different from those underlying Hubble.

In Compton scattering, a gamma ray interacts with an electron in such a way that the electron recoils and the gamma ray is deflected, with the angle of deflection related to the energy transferred from the incoming gamma ray to the recoil electron. Historically, the Compton effect was influential in showing the particulate nature of electromagnetic radiation. In the CGRO, it was put to use in the design of instruments.[3] Here I will discuss only two of the four instruments aboard CGRO, and only one in any detail.[4]

The Imaging Compton Telescope (COMPTEL), shown diagrammatically in figure 3.2, had two levels of detectors separated by 1.5 meters. The upper detector consisted of 7 cylindrical modules of liquid scintillator. Each module was 27.6 cm in diameter, 8.5 cm thick, and viewed by eight photomultiplier tubes. The total area of the upper detector was approximately 4,188 square cm. The lower detector consisted of fourteen cylindrical sodium iodide blocks 7.5 cm thick and 28 cm in diameter. Each block of sodium iodide was viewed from below by seven photomultiplier tubes. The total geometrical area of the lower detector was 8,620 square cm. In addition, each detector was surrounded by a shield consisting of a 1.5 cm-thick dome of plastic scintillator, viewed by twenty-four photomultiplier tubes. The purpose of

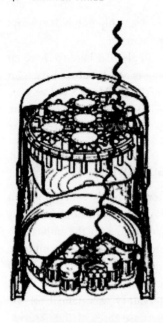

Figure 3.2 Diagram of the COMPTEL instrument. (Image courtesy of NASA.)

the shield was to detect stray charged particles and, using anticoincidence circuitry, reject their interactions with the detectors.

COMPTEL was efficient at detecting gamma rays in the 1 to 30 MeV range. An incident gamma ray was Compton scattered in the first detector, the recoiling electron producing a scintillation whose intensity was measured by the photomultiplier tubes in the relevant module. The scattered gamma ray, now at a much reduced energy, was again Compton scattered in the second detector, also producing a scintillation measured by the relevant photomultiplier tubes. The total energy of the incoming gamma ray, then, was assumed to be roughly the sum of the energies measured in the two detectors, accurate to within 5 or 10 percent. By comparing the energies detected individually by the seven or eight photomultiplier tubes in the relevant modules, and knowing the relative locations of the modules, the direction of an incoming gamma ray can be estimated to within a few degrees. A number of events will show a clustering in a given direction, thus identifying the spatial location of sources of the gamma rays. To ensure that the two measurements did indeed represent the same gamma ray, the timing of the two events had to correspond to the expected time of flight of a gamma ray over the meter and a half between detectors. Additionally, of course, the shield photomultiplier tubes must not, in this interval of time, indicate passage of a stray charged particle through the apparatus.[5]

Plate 6 (in the color insert) reproduces a dramatic COMPTEL image of the center of the Milky Way Galaxy. This image was constructed from data collected over a period of five years restricted to gamma rays measured at

1.8 MeV (plus or minus 5 to 10 percent). Why 1.8 MeV? Well, that is the energy of gamma rays produced by the decay of a radioactive isotope of aluminum, aluminum 26, into magnesium 26. It has been theorized that heavy elements like Al 26 are produced in the cores of very massive stars such as those existing in the center of galaxies. Since there are unlikely to be significant other sources of 1.8 MeV gamma rays, this image confirms long-held theories of star formation. In this false color image, regions of no 1.8 MeV emission are dark, while blue, green, red, and yellow indicate regions of increasingly intense emissions. A surprising feature of this image is the apparent "halo" of 1.8 MeV gamma rays, shown in blue in the image, surrounding the galactic center. The existence of this halo was unknown before this image was produced.

Plate 7 (in the color insert) is also an image of the center of the Milky Way produced by another of CGRO's instruments, the Oriented Scintillation Spectrometer Experiment (OSSE). I will not describe the operation of this instrument in any detail. Its region of maximum sensitivity, .1 to 10 MeV, was slightly lower than that of COMPTEL. For the observations shown in plate 7, its energy range was restricted to 0.51 MeV (plus or minus 5 to 10 percent). Why 0.51 MeV? Because that is the energy of gamma rays produced by electron-positron annihilation. Earlier surveys had indicated a particular intensity of such gamma rays. What was surprising about this image was that it indicated the existence of a quite intense plume of positrons extending asymmetrically at right angles to the plane of the galaxy. At the time this image was released to the public, there was no accepted explanation for the existence of such a plume. Much speculation centered on the possibility of it being produced somehow by a large black hole at the center of the galaxy. But, at the time, nobody knew exactly how.[6]

My purpose in presenting these two very different images of what is theoretically the same object is to emphasize the conclusion that it is at best misleading to refer to these simply as images of the center of the Milky Way. The character of these images owes as much to the instruments that produced them as to the nature of the Milky Way. What we have here is the Milky Way as viewed from the perspective of COMPTEL and the Milky Way as viewed from the perspective of OSSE. These instruments share the perspective of seeing only gamma rays. But they differ in their sensitivity to different energies of gamma rays, and these two images differ in the specific energies to which they are restricted. Additionally, the character of the images differs as a result of being produced by significantly different processes, both physically different and computationally different. Finally, they have different error characteristics, which means they differ in how they group together gamma rays that are theoretically distinct.

In the previous chapter I emphasized the comparative study of color vision, which compares color vision in humans with color vision in other animals. There I argued that the comparative study of color vision supports a perspectival (relational) account of color perception. I would now like to suggest that what I have just presented contains aspects of a comparative study of radiation detection in general. I have compared the ability of several different instruments to detect radiation in various regions of the electromagnetic spectrum. Here I advance the parallel argument that the comparative study of electromagnetic detectors supports a perspectival understanding of radiation detection and of scientific observation more generally.

Humans and various other electromagnetic detectors respond differently to different electromagnetic spectra. Moreover, humans and various other electromagnetic detectors may face the same spectrum of electromagnetic radiation and yet have different responses to it. In all cases, the response of any particular detector, including a human, is a function of *both* the character of the particular electromagnetic spectrum encountered and the character of the detector. Each detector views the electromagnetic world from its own *perspective*. Every observation is perspectival in this sense.[7]

Conclusions within Perspectives

One way of understanding my claim that scientific observation is perspectival is to say that claims about what is observed cannot be *detached* from the means of observation. Observation does not simply reveal the intensity and distribution of gamma rays coming from the center of the Milky Way; it reveals the intensity and distribution of gamma rays *as indicated by COMPTEL or OSSE or* . . .

It might be thought I am saying that, in giving reports of observations, one should not go beyond the data. One should just report the data and not draw conclusions. But this is *not* my view. The images presented in plates 6 and 7 *are* conclusions. These images present a picture that is continuous, or at least very fine-grained. The actual data cannot be that fine-grained. The data are made up of individual events recorded in various detectors at different times and processed by various physical and computational means. The images are constructed using those data, but go beyond the data. To invoke an idea to be discussed in chapter 4, the images are *models of the data*, not the data themselves.

And how does one get from the actual data to models of the data? By employing standard scientific methods adapted to the circumstances: correcting for background noise, calibrating components, repeating measurements, performing statistical analyses, and so on. This is just what everyone would call sound scientific practice. Nevertheless, this practice, I claim, remains *internal* to the relevant perspective. It does not get one beyond the perspective of the instrumentation employed. One still ends up, for example, with a model of the data produced by COMPTEL.

But surely, it will be objected, scientists draw conclusions going beyond their instrumentation. Indeed they do. But they do so only by moving to a broader *theoretical* perspective. So I shall argue in chapter 4. First, however, I would like to continue the discussion of observational perspectives in the context of a very different science, neuroscience. This will support the conclusion that the lessons learned in the study of color vision carry across the sciences.

Imaging the Brain

By an act of the United States Congress, the 1990s were designated "The Decade of the Brain." No doubt someone has enthusiastically labeled the twenty-first century "The Century of the Brain." Exaggerations aside, during the first half of the twenty-first century, the cognitive sciences, particularly the neurosciences, are likely to occupy the sort of position among all the sciences that was held by physics in the first half of the twentieth century and by biology, especially molecular biology, in the second half.

One major factor in the recent explosion of interest in the neurosciences is the development of powerful new imaging technologies for investigating the structure and function of the living brain.[8] It may not be an exaggeration to say that the neurosciences are currently being driven more by imaging technology than anything else, including theory. Figure 3.3 gives a rough chronology of the introduction of the major imaging technologies along with an indication of their optimal two-dimensional spatial resolution.

Single Photon Emission Computed Tomography
(SPECT). 1950-1960. 10 sq. mm.

Computer Assisted Tomography (CAT, CT)
Early 1970s. 5 sq. mm.

Positron Emission Tomography (PET).
Mid-1970s. 5 sq. mm.

Magnetic Resonance Imaging (MRI).
Early 1980s. 1 sq. mm.

Functional Magnetic Resonance Imaging (fMRI).
Early 1990s. 1 sq. mm.

Figure 3.3 Chronology and resolution of imaging technologies.

The word *tomo* is Greek for "slice." Originally, tomography was the study of a slice of tissue using a microscope. Many contemporary imaging techniques continue this tradition by recording information in two-dimensional slices, though now the slices are purely geometrical rather than physical. As a point of reference, plate 8 (in the color insert) reproduces a color photograph of a stained and frozen axial slice of a human brain. One must be careful not to think that this picture represents a section of the brain "as it really is." This is at best how such a slice appears from the perspective of a normal human trichromat, mediated by stains and color film designed to match the chromatic sensitivity of the human visual system.

Computer Assisted Tomography (CAT)

We begin with computer assisted tomography, sometimes called simply computed tomography (CT). The phrase "CAT scan" has entered the general vocabulary. So just what is a CAT scan? CAT is an elaborate form of X-ray imaging. From ordinary X-ray photographs, we understand that X-rays are attenuated as they pass through various materials; the denser the material, the greater the attenuation. So bones show up as roughly white on negative film, with soft tissue yielding varying shades of gray. In CAT, a whole array of electronic detectors records the intensity of transmitted X-rays in a narrow band along lines from the source to detector. This is done at a range of angles all around the head of a subject, as shown in figure 3.4. That is the "scan" part of "CAT scan." From the many resulting one-dimensional recordings of relative attenuation, a computer with appropriate software can reconstruct the two-dimensional structure of a slice of tissue without physically slicing it. The result is a lower resolution version of what one would get if one in fact sliced the brain tissue and then took standard X-ray pictures through the slices.[9] Plate 9 (in the color insert) provides a representative example.

The most important general feature of the perspective provided by computer assisted tomography is that it is *structural*. It provides us with

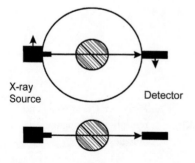

Figure 3.4 Design of instrument for CAT.

information about the *structure* of the brain. In particular, it provides no information about the *function* of any of the structures it reveals. From a representational point of view, this is what most strongly distinguishes computer assisted tomography from some other forms of brain imaging.[10]

Positron Emission Tomography (PET)

Let us turn now to a form of neuroimaging that provides a *functional* perspective on the brain, positron emission tomography, commonly known by its acronym, PET, as in "PET scan." PET works with gamma rays, which are even more energetic than typical X-rays. But rather than the radiation originating *outside* the brain, it originates *inside* and radiates out. In more technical terms, PET works by *emission* rather than *transmission* of ionizing radiation. The source of the radiation is one of a half dozen or so short-lived radioactive isotopes attached to biologically active molecules that are typically injected into the bloodstream of the subject.

The isotopes decay by emitting a positron and a neutrino, as shown in figure 3.5. The neutrino escapes, leaving no trace, but the positron very soon collides with an electron, resulting in the mutual decay of the electron-positron pair into two gamma rays of energy 511 keV each. To satisfy conservation of momentum, these two gamma rays must travel in exactly opposite directions. In classic PET machines, as shown schematically in figure 3.6, the subject is surrounded with an array of small gamma ray detectors wired so as to record only very close coincidences in received gamma rays. When a coincidence is detected, it is then assumed that the gamma rays originated somewhere along the line between the two detectors in the plane determined by the location of the whole array of detectors. If there is a concentration of the radioactive molecules in a particular location, the detectors will record many lines crossing at that location. A computer program then processes all these readings to produce a two-dimensional picture of the resulting slice of

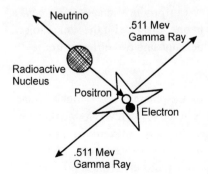

Figure 3.5 The physics behind PET.

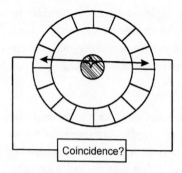

Coincidence?

Figure 3.6 A schematic representation of a PET machine.

the brain with relative intensities, coded by color, indicating locations where the injected substance is more concentrated. Red coloring indicates the highest level of concentration, and so on down the color spectrum. Plate 10 (in the color insert) exhibits a representative PET image.

There are several dozen different neuroactive substances that can be tagged with positron emitting isotopes. Which substance is used determines what functions can be localized in the brain. The largest number of such substances are involved in metabolism or biosynthesis, but some are neurotransmitters or neuroreceptors, thus indicating locations where more specialized functions are performed. So, not only does PET provide a functional perspective on living brains, it provides a number of different functional perspectives depending on the nature of the injected substance.

Given the great differences between the structural and functional perspectives, an obvious strategy is to combine them, producing an image with functional information superimposed upon a structural background. This is much more difficult than it sounds if the two images for a single subject must be produced on entirely different machines in different places and at different times. Just getting the two images to the same scale is difficult. But it is done. It is one of the many advantages of the most recent magnetic resonance imaging instruments that they can produce images from *both* a structural and functional perspective in one session with the same subject, thus eliminating most problems of superimposing two different images.

Magnetic Resonance Imaging (MRI)

Magnetic resonance imaging is based on the phenomenon of nuclear magnetic resonance discovered by physicists in the 1940s. It was found that people in biomedical contexts reacted negatively to the word *nuclear*, so, in this context, the phrase got shortened to simply "magnetic resonance." The phenomenon is complex, depending on the quantum mechanical properties of nuclei. For my purposes here, a simplified account will suffice. I will consider only the simplest atom, hydrogen, whose nucleus consists of a single proton, and ignore the accompanying electron.[11]

A proton has an intrinsic spin, like a top. Because it is charged, the spinning motion produces a small magnetic field. So a proton acts like a small magnet. In ordinary matter, the orientation of individual protons is random. If subjected to an external uniform magnetic field however, protons, like magnets, tend to line up with the field. Unlike macroscopic magnets, protons, being of quantum dimensions, can take only two different orientations: either with the field or against the field. There is an energy difference between these two states: going against the field requires higher energy. But, again being of quantum dimensions, the protons move randomly from one orientation to another. On the average, however, there is a slight bias in favor of the lower energy state, roughly ten in a million at room temperature in a moderate (1.5 Tesla) magnetic field. But since there may be roughly 10^{23} protons in a cubic centimeter of matter, that leaves an excess of roughly 10^{18} protons, enough to produce a measurable *macroscopic* induced magnetic field called the *longitudinal field*, M_1. Figure 3.7 shows the microscopic longitudinal field for an individual proton in its low energy state.

When the imposed magnetic field is turned on, it takes some time, called the *longitudinal relaxation time*, or T_1, for the macroscopic longitudinal field, M_1, to reach equilibrium. The precise energy needed to bump protons into alignment with the imposed field comes from collisions with other atoms in its surroundings. The efficiency of this energy transfer thus depends on the molecular makeup of the material in question. For typical biological materials, these relaxation times vary from tens of milliseconds to several seconds. If one could localize the source of the induced longitudinal field, one would therefore have an indication of the specific materials at that position, and thus information to construct a structural representation. More useful signals, however, can be obtained by taking advantage of even more complex quantum mechanical interactions.

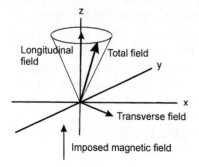

Figure 3.7 The longitudinal NMR field.

Like a spinning top that has been bumped, a proton in an imposed magnetic field processes around its average axis. So there is another *component* of the field, the *transverse field*, shown in figure 3.7, that rotates around the axis. The macroscopic sum of many such individual fields is called, simply, the transverse field, M_2. The natural *frequency* of rotation, known as the Larmor frequency, is a fixed function of the strength of the imposed magnetic field and lies in the range of radio frequencies (RF).[12] But because there is no coordination among individual protons, the transverse component of the overall resultant field at the macroscopic level cancels out. This can be changed, however, by imposing a radio frequency signal operating at exactly the natural frequency of rotation of the protons, thus producing a *resonance* between this natural frequency and the imposed radio frequency signal. Here is the "resonance" in "magnetic resonance imaging." If the amplitude and duration of a pulse of this RF radiation are just right, it may suppress the longitudinal component of the induced field, leaving only the transverse field as the total field. The imposed RF field also forces all the individual transverse fields of all the protons in the designated volume to rotate together, in phase. The sum of these rotating fields produces a measurable macroscopic radio signal at the resonance frequency. One requires only an appropriate "antenna" to detect this field.

How is all this useful for producing images of the brain? Once the external pulse of RF radiation is finished, the transverse rotation of individual protons begins to go out of phase because of interactions with other atoms. So the intensity of the emitted RF radiation decays. The important fact is that the *rate* of this decay is sensitive to the specific atomic environment of the protons, which, remember, are nuclei of hydrogen atoms. So different substances exhibit different decay times, called T_2, for the emitted *transverse* RF radiation. If one can measure these differences, different kinds of tissues, even different chemicals, can be distinguished. Actually making such measurements requires a further quantum mechanical trick, invoking what is known as "spin echo," which I will omit here.

How such measurements can be *localized* to a particular cubic millimeter within a brain, however, is worth examining. The key is that the resonance frequency of protons depends on the strength of the imposed static magnetic field. Introducing an appropriate *gradient* in this field, as shown in figure 3.8, produces protons with *different* resonance frequencies as one moves along the y-axis. A narrow bandwidth of the imposed RF radiation will then pick out a thin volume element parallel to the x-z plane where the protons will be in resonance with the RF field. Thus, by varying the frequency of the imposed RF field, one can in effect sweep

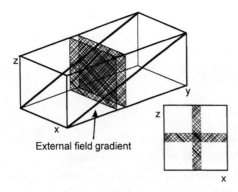

External field gradient

Figure 3.8 Localizing the source of the emitted radio frequency signal.

through all such volume elements along the *y*-axis. So one can measure the decaying induced transverse field from just one such volume element at a time.

In principle, one can imagine this process being repeated with the impressed magnetic field gradient being along the *x*-axis and then along the *z*-axis. The intersection of any three such slices would thus pick out a small cube of material. Combining the three signals in an appropriate manner would yield a unique relaxation time, T_2, for each small cube of material. Putting together measurements from a two-dimensional array of such cubes yields an image like that shown in plate 11 (in the color insert). In practice, such a method would be too slow to use effectively with living and breathing subjects. So modifications known by such names as FISP (fast imaging with steady-state precession), RARE (rapid acquisition with refocused echoes), and EPI (echo-planar imaging) have been devised to speed up data acquisition.

Functional Magnetic Resonance Imaging (fMRI)

I introduced this discussion of MRI by noting that it made possible the acquisition of both structural and functional perspectives with one instrument in one experimental session. A now-standard way of obtaining functional images is to take advantage of the fact that there is a small difference in the magnetic susceptibility of oxygenated blood (oxyhemoglobin) versus deoxygenated blood (deoxyhemoglobin). Since 1990, MRI techniques have been developed that are sensitive enough, and fast enough, to detect locations in which one or the other blood type predominates. And oxygen utilization, of course, correlates with brain activity. Thus, regions of higher and lower activity can be located. Finally, one can artificially produce differences in magnetic susceptibility by introducing various substances with known magnetic and physiological properties into the bloodstream. Relative differences in the concentration of these substances can then be detected, indicating regions where the corresponding functions are taking place.[13] Plate 12 (in the color insert) is a representative fMRI image.

Instrumental Perspectives

Now, following my account of imaging in astronomy, I would like to say that the images produced by various brain imaging technologies are perspectival. Computer assisted tomography, for example, provides a *perspective* on the structure of the brain. Many factors besides the structure of the brain contribute to this perspective; in particular, the nature of X-rays and their interaction with various kinds of tissue, the design and operation of the detectors, and the elaborate computer program that converts relative linear intensity data into two-dimensional black-and-white images. Different brain tissues that attenuate X-rays to about the same extent show up the same on a CAT scan. Differences in intensities of X-rays less than the resolving power of the detectors are not represented in the resulting image. In sum, the images CAT scanners produce should not be thought of as simple pictures of the brain. They are images of the brain *as produced by the process of computer assisted tomography.*

These points are, if anything, even stronger in the case of MRI. The image that is produced is the result of complex atomic interactions understandable only in quantum mechanical terms. The RF signal detected is produced by induced nuclear magnetic resonance, again a quantum mechanical phenomenon. The choices regarding exactly which parameters to measure, T_1 or T_2 or some combination, and how to process them in the final image production are strongly influenced by considerations that often trade off speed of data acquisition against sensitivity to differences in tissue composition. There is no one "right" or "best" way to produce MR images. There are many ways, all with their own virtues and shortcomings relative to various investigative aims. MRI, in particular, makes it abundantly clear that not only is scientific observation perspectival, but also that there are multiple perspectives from which one must choose and no "objectively" correct choice. A lot depends on the goals of the investigation at hand.

In sum, scientific observation does not simply produce images of the brain. One has images *as produced by CAT or MRI or so forth.* One cannot detach the description of the image from the perspective from which it was produced. This goes even for the color photograph of a frozen brain section reproduced in plate 8.

The above, of course, are claims about the scientific status of observational claims from the perspective of a scientific account of the activity of doing science. I am not trying to legislate how scientists should talk in their everyday scientific pursuits. When scientists put up images on a screen at a professional meeting, they may simply say, "As you see in this image, ..." But the experts in the audience know, or will already have been told, how the image was produced. For the most part, they know the virtues and limitations of various imaging techniques and are rarely tempted to say, or even think, that this is how the brain somehow "really looks."

The Compatibility of Instrumental Perspectives

In chapter 2 I claimed that we expect different visual systems viewing the same scene from the same location to be compatible because we operate with the methodological presumption that the total electromagnetic field at any place is unique. Differences in images produced must be due to differences in the makeup of the different visual systems. There is nothing at all contradictory about different systems producing different images with the same input.

This is not to say that different instruments might not yield results that *appear* to conflict. Two different gamma ray detectors may overlap in the range of energies of gamma rays to which they are sensitive. So it is possible, for example, that one instrument might indicate that, over a given period of time, there is a considerable flux of gamma rays of energy 10 MeV coming from a well-defined source, while the other indicates hardly any flux at that energy during that period of time. The relevant group of scientists confronted with this situation would draw the conclusion that one or the other instrument is malfunctioning and proceed to try to figure out what had gone wrong. They would not accept the result as simply a curiosity of nature. This behavior would be in accord with the methodological principle of proceeding as though nature has a unique causal structure. Of course, one does not expect working scientists explicitly to invoke anything so grandiose. That is just how I recommend we understand their activities.

Overlapping Instrumental Perspectives

It is a commonplace that there can be many observational perspectives of the same objects. The Milky Way, for example, can be perceived from an unaided white light perspective by normal humans, provided they get far enough away from the ambient light of modern cities. Optical telescopes, as well as more exotic instruments, such as gamma ray telescopes like COMPTEL and OSSE, provide other perspectives. Is this not good evidence that there is something "objectively" there? Indeed, this is good evidence that there is *something* there, but this need not be understood as knowledge in an "absolute objectivist" sense.

The simple but fundamental point is that to be an object detected in several different perspectives is not to be detected in no perspective whatsoever. All observational claims made about the object are made in some perspective or other. Before the seventeenth century, the Milky Way, as part of a commonsense perspective on the world, was perceived using human eyes simply as a broad band of light extending across the night sky. From the perspective of Galileo's roughly thirty power telescopes, it was perceived as being made up of a very large number of individual stars. But this was a change in perspective, not a move from a mere perspective to objectivist truth.[14] Moreover, that what he was seeing through his telescope was the same object he could see with his unaided eyes was a claim he could make within his own expanding commonsense perspective on the world. He needed only to look at the Milky Way and point his telescope in that direction. Within the perspective created by large reflecting telescopes built in the twentieth century, we can see some of what Galileo thought were just other stars as being distinct galaxies composed, in turn, of millions of stars. Again, because we can experience the Milky Way from all these different perspectives, we know it is the same object, just experienced from different perspectives. And so on to observations made with instruments such as COMPTEL and OSSE. Similar remarks apply to the object Galen identified as the human brain.

The existence of overlapping perspectives is not merely of interpretive importance. It also plays an important role in actual scientific investigations. In astronomy, to give just one example, finding an optical light source corresponding to a radio source can provide good evidence that the source of the radio signals has been correctly located. It also gives confidence that the radio signal was not merely an artifact of the radio telescope itself. This important methodological strategy can, however, be well understood in perspectival terms.

CHAPTER FOUR
SCIENTIFIC THEORIZING

Introduction

I have argued that both human color vision and scientific instruments are perspectival, and in several senses. In a broad sense, they are perspectival in that they interact with only restricted aspects of the world, visible light, gamma rays, or the atomic structure of brain tissue. But these instruments are perspectival also within their more general perspectives. They process inputs from the environment in ways peculiar to their own physical makeup, ways that render these inputs similar or different not just according to features of the inputs themselves, but also according to features of the instrument.

It is now time to face the obvious question noted at the end of chapter 2. On what basis do I make all these claims about color vision and scientific instruments? The short answer is, on the basis of various scientific theories; theories of color vision, of electromagnetism, even quantum theory. But this answer immediately raises a further question: Are not scientific theories to be understood in an objectivist framework? If so, I remain an objectivist at the level of theory even if theory itself shows us that observation is perspectival. My reply is that theoretical claims are also perspectival.[1]

The basic idea is that conception is a lot like perception, or, that theorizing is a lot like observing. More specifically, in creating theories, I will argue, scientists create perspectives within which to conceive of aspects of the world. To make this argument, I must first present a conception of theorizing.[2]

Representing

Many commentators on the nature of science, whether from an objectivist or constructivist point of view, assume that the theoretical claims of scientists are expressed primarily in declarative sentences, perhaps in a particular form, such as the form of laws of nature. This could be expressed by saying that scientists represent the world primarily in linguistic terms, and this leads, in turn, to a focus on representation understood as a two-place relationship between linguistic entities and the world. This view of representation is so deeply entrenched, particularly in philosophy and the philosophy of science, that it would be hopeless for me here to mount a direct attack on this view. I will therefore concentrate on presenting an alternative point of view.

I suggest shifting the focus to scientific *practice*, which implies that we should begin with the practice of *representing*.[3] If we think of representing as a relationship, it should be a relationship with more than two components. One component should be the agents, the scientists who do the representing. Because scientists are *intentional* agents with goals and purposes, I propose explicitly to provide a space for purposes in my understanding of representational practices in science. So we are looking at a relationship with roughly the following form: S uses X to represent W for purposes P. Here S can be an individual scientist, a scientific group, or a larger scientific community. W is an aspect of the real world. So, more informally, the relationship to be investigated has the form: Scientists use X to represent some aspect of the world for specific purposes. The question now is, What are the values of the variable, X?

Focusing on scientific practice, one quickly realizes that X can be many things, including, of course, words and equations, but also, for example, diagrams, graphs, photographs, and, increasingly, computer-generated images.[4] Here, however, I wish to begin by focusing on the traditional medium of scientific representation, the scientific *theory*. So the questions become, What are scientific theories? and How are theories used to represent the world?

Theories

The assumption that scientific representation is to be understood as a two-place relationship between statements and the world goes along with the view that scientific theories are sets of statements. A focus on the activity of representing fits more comfortably within a model-based understanding of scientific theories. Figure 4.1 provides an abstract picture of such a view of theories.

In this picture, scientists generate models using principles and specific conditions.[5] The attempt to apply models to the world generates hypotheses about the fit of specific models to particular things in the world. Judgments of fit are mediated by models of data generated by applying techniques of

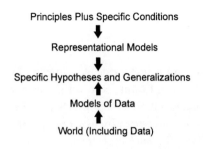

Principles Plus Specific Conditions

Representational Models

Specific Hypotheses and Generalizations

Models of Data

World (Including Data)

Figure 4.1 Overview of a model-based account of theories.

data analysis to actual observations. Specific hypotheses may then be generalized across previously designated classes of objects.

Principles

In some sciences, models are constructed according to explicitly formulated principles. Physics is especially rich in such principles: Newton's principles of mechanics; Maxwell's principles of electrodynamics; the principles of thermodynamics; the principles of relativity; and the principles of quantum mechanics. But evolutionary biology also has its principle of natural selection and economics boasts of various equilibrium principles.

What I am here calling principles have often been interpreted by empiricist philosophers as empirical laws, that is, generalizations that are both universal and true. My view (Giere 1988; 1999a), which I share with Nancy Cartwright (1983, 1999), Paul Teller (2001, 2004a), and some others, is that, if understood as universal generalizations, the resulting statements turn out to be either vacuously true or else false, and known to be so. The remaining problem is how otherwise to characterize these principles.

I think it is best not to regard principles themselves as vehicles for making empirical claims. Newton's three laws of motion, for example, refer to quantities called force and mass and relate these to quantities previously understood: position, velocity, and acceleration. But the principles do not themselves tell us in more specific terms what might count as a force or a mass. So we do not know where in the world to look to see whether or not the laws apply. One can give a similar account of the evolutionary principles of variation, selection, and transmission.

If we insist on regarding the linguistic formulations of principles as genuine statements, we have to find something that they describe, something to which they refer. The best candidate I know for this role would be a highly abstract object, an object that by definition exhibits all and only the characteristics specified in the principles.[6] So the linguistic statements of the principles are true of this abstract object, though in a fairly trivial way. I would now summarize this view by saying that this abstract object is a very general model whose initial function is to characterize relationships among the elements of the model.[7]

The most important thing to understand is how principles function in representational practice. Their function here, I think, is to act as general templates for the construction of more specific models, which are also abstract objects. Thus, to the principles one adds what I am here calling "specific conditions," the result being a more specific, but still abstract, object. The principles thus help both to shape and also to constrain the structure of these more specific models. To take a canonical example, adding the condition that $F = -kx$ to Newton's principles yields a general model for a simple harmonic oscillator, where x is the displacement from an equilibrium position.

With this model we are still some distance from any empirical claims. This model could be applied, for example, to a pendulum with small amplitude, a mass hanging from a spring, the end of a cantilevered beam, or a diatomic molecule. But even specifying that x is the displacement of a mass on a spring does not yet get us to an empirical application. We still have only an abstract model of a mass on a spring. One way of getting down to an actual empirical claim would be to designate a particular real mass on a spring. This would result in a maximally specific (but still abstract) model in which every relevant element of the model is identified with some aspect of a system in the real world. It could then be empirically determined whether the motion of that particular mass on that particular spring agrees with the motion calculated for the abstract mass in the model.

The movement from principles to models involves two activities that are too easily conflated. One I call "interpretation." The principles of Newtonian mechanics, for example, help to interpret the terms *force* and *mass* within a Newtonian perspective by showing their relationships with the terms *position*, *velocity*, and *acceleration*. The principles do not, by themselves, determine the interpretation. How that is done is a question for a general theory of language, now a task for the cognitive sciences. I shall say no more about that here.[8] A second activity involved in deploying models is the "identification" of specific things in the world with elements of a model. If all relevant elements of a model are so identified, the model is maximally specific. Again, *how* scientists perform this task seems to me not a problem peculiar to a theory of science, but to the understanding of language use more generally.[9]

Representational Models

At first sight, the things that, within the sciences, are commonly called "models" seem to form a quite heterogeneous class including physical

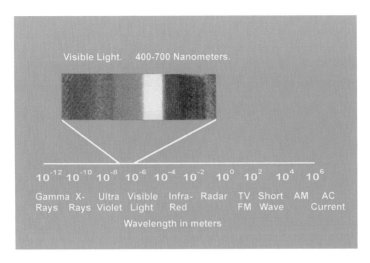

Plate 1 The electromagnetic spectrum.

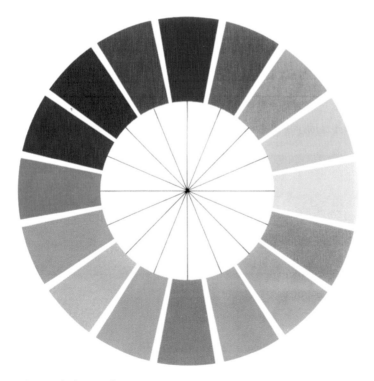

Plate 2 The hue circle.

Plate 3 Comparison of dichromatic and trichromatic views of the same scene.

Plate 4 The Trifid Nebula. (Image courtesy of Anglo-Australian Observatory/ David Malin Images.)

Plate 5 Hubble (ACS) photo of January, 2003. (Image courtesy of NASA.)

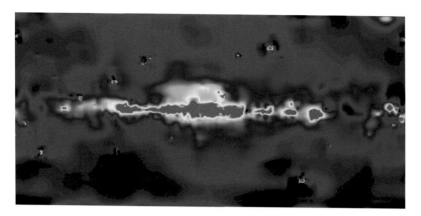

Plate 6 COMPTEL image. (Image courtesy of NASA.)

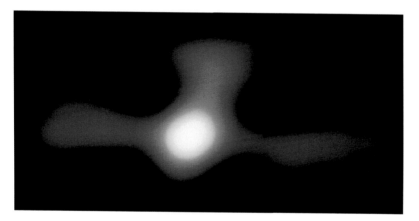

Plate 7 OSSE image. (Image courtesy of NASA.)

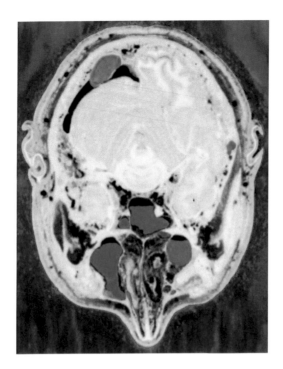

Plate 8 Color photograph of a stained axial slice of human brain. (Image courtesy of NIH.)

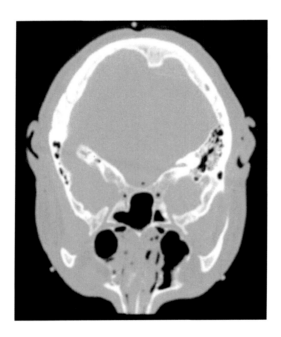

Plate 9 A representative CAT scan. (Image courtesy of NIH.)

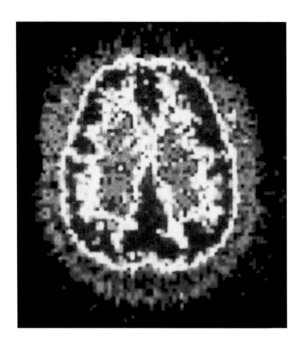

Plate 10 A representative PET scan. (Image courtesy of NIH.)

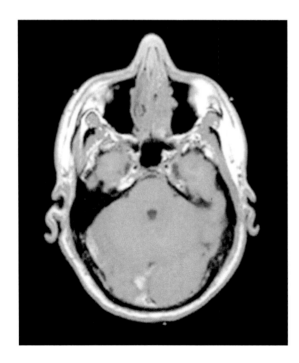

Plate 11 A representative MRI image. (Image courtesy of NIH.)

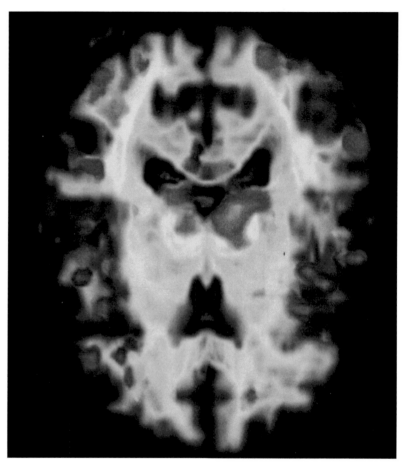

Plate 12 A representative fMRI image. (Image courtesy of the University of Minnesota.)

models, scale models, analogue models, and mathematical models, to name just a few. Thus we have, for example, Watson and Crick's original sheet metal and cardboard model of DNA, Rutherford's solar system model of atoms, the Bohr model of the atom, and the de Sitter model of space-time. There are also equilibrium models in economics and drift models in evolutionary biology. I think it is possible to understand scientific models in a way that usefully encompasses much of this heterogeneity.

Following the general schema of figure 4.1, models in advanced sciences such as physics and biology should be abstract objects constructed in conformity with appropriate general principles and specific conditions. One might think of them as artful specifications of the very abstract models defined by the principles.[10] What is special about these models is that they are designed so that elements of the model can be identified with (or coordinated with) features of the real world. This is what makes it possible to use models to *represent* aspects of the world. So here, finally, we have a candidate for the X in the general scheme for representation with which we started: Scientists use *models* to represent aspects of the world for various purposes. On this view, it is models that are the primary (though not the only) representational tools in the sciences. In order to distinguish these models from the more abstract objects defined by principles, I may sometimes refer to them as "representational models." They are designed for use in representing aspects of the world. The abstract objects defined by scientific principles are, on my view, not intended directly to represent the world.[11]

My view of models might be described as an "agent-based" account of models. This view of models supports a model-based understanding of scientific theories. Later in this chapter I will sketch a similarly agent-based account of the empirical testing of representational models against instrumental perspectives. Thus, over all, the picture of science that emerges is an agent-based picture. In the following chapter, I will argue that the fundamental agents are *human* agents.

Similarity

How do scientists use models to represent aspects of the world? What is it about representational models that makes it possible to use them in this way? One way, perhaps the most important way, but probably not the only way, is by exploiting possible *similarities* between a model and that aspect of the world it is being used to represent. Note that I am not saying that the model itself represents an aspect of the world because it is similar to that aspect. There is no such simple representational relationship.[12] Anything is similar to anything else in countless respects, but surely not everything by itself represents something else. It is not the model that is doing the representing; it is the scientist using the model who is doing the representing.[13] One way scientists do this is by picking out some specific features of the model that are then claimed to be similar in some specific respect to features of the designated real system. It is the possibility of specifying such similarities that makes possible the use of the model to represent the real system in this way.

It is now clear that the above account of using abstract models to represent real systems applies as well to the use of physical models. To return to a previous example, it was particular similarities in physical structure that made possible Watson and Crick's use of their sheet metal and cardboard model to represent the structure of DNA. They clearly were not saying that DNA is similar to this model with respect to being composed of metal and cardboard. Part of using a model to represent some aspect of the world is being able to pick out the relevantly similar features. Another part of using a model to represent something is having some reasonable idea of how good a fit might be expected. The angles in Watson and Crick's model used to represent bonding angles in DNA were not exactly the bonding angles later determined for samples of DNA. But no one doubted they were close enough to conclude that DNA has a double helical structure. Moreover, the angles in the model were somewhat adjustable, and so could be made better to fit the angles in DNA as more precisely determined by later experiments.

Representing aspects of real systems in this way does not require the existence of a general measure of similarity between models and real systems. I doubt that there exists any uniquely justifiable measure of this type. It is sufficient that a reasonable, even if not unique, measure can be specified for any particular claimed similarity. Thus, in the example of the mass hanging from a spring, using the mathematical characterization of the model, one can calculate the period of the oscillation as a function of k/m. Measuring the value of this parameter for the real spring system, one can then determine how close the measured value of the period is to the value calculated for the model. One might determine, for example, that the observed period, T_O, measured in tenths of a second, is within 5 percent of the theoretically calculated value. Whether this would count as "similar enough" would depend on the purposes for which the model is being applied. That is a function of the context and not merely a relationship between the model and the system to which it is applied.

Similarity and Truth

Strictly speaking, it makes no sense to call a model itself true or false. A model is not the kind of thing that could have a truth value. In fact, models are more like predicates than like statements.[14] So calling a model true would be like calling a predicate true, which is nonsense. On the other hand, we can correctly say that a predicate is "true of" some particular thing or "applies truly" to that thing. Likewise, one might be tempted to say that a *model* is "true of" some particular thing or "applies truly" to that thing. This is acceptable as an ordinary way of speaking, but one must be careful not to read too much into this way of speaking. To say a model is "true of" a particular real system in the world is to say no more than it "fits" that system or "applies to" that system.

To say a model fits a particular system does, however, imply various individual statements about that system. To claim that a particular mass on a spring fits the standard model implies that there is a particular value of k, the spring constant, a particular value of T, the period of oscillation, and so on. These claims look to be straightforwardly either true or false. But claims about the values of particular parameters for a specific real system do not exhaust the content of the overall claim that the model fits that system. The model also specifies the dynamical relationships among variables, for example, the position as a function of time and the relationship between the period of oscillation and the spring constant. If we add statements describing all of these relationships, then we will have indeed exhausted the content of the general claim that the model "fits," or is "true of," the real system. But we still need a general notion such as "fit," "applies to," or "being true of" both to make general claims about the fit of particular models to particular systems and to make even more general claims about the fit of unspecified models to unspecified systems. The question is whether we need a separate notion of "similarity" between models and real systems.

In ordinary life, truth claims are quite flexible along several dimensions. If we say Spike is tall, that most likely means that Spike is over six feet tall though not over seven feet tall, and probably not even over six feet six inches tall. Ordinary discourse allows such indeterminacy; indeed, ordinary discourse requires it if ordinary linguistic transactions are to proceed smoothly and efficiently. Ordinary discourse also allows relativization to subcategories. Thus, one could affirm that Spike is tall for an African American man but still rather short for a center on a professional basketball team. Of course there are also ordinary circumstances that require more precision, such as the published statistics on players for a particular team. This list might say that Spike is six feet seven inches tall, though probably not that he is six feet seven and one-sixteenth inches tall.

The demand for precision is in general greater in the sciences than in everyday life, although even here how much precision is required depends on the context. What is different in the sciences, I think, is the conscious awareness that models cannot be expected to fit their intended subjects with perfect precision. This realization is, of course, strongest in fields for which there typically exist quantitative models, but is present even when most models are relatively qualitative. It is partly to emphasize this aspect of the use of models in the sciences that I have always preferred to think of the general relationship between models and the systems they model in terms of "similarity" rather than some variant on truth such as "true of." Talk of similarity makes explicit the fact that one does not expect there to be a perfect fit between models and the world. It also leaves open what specific relationships among elements of the model are in question and what measure one will use to specify how closely these relationships track the corresponding relationships within the real system. There is most likely no single measure that is uniquely best for determining goodness of fit for every variable in a model, or even just for those variables corresponding to elements of a real system that can be measured experimentally.[15]

My experience is that scientists with objectivist intuitions, particularly those concerned with fundamental physics, tend to think it possible that there could be fundamental models that fit the world exactly. This must currently be an unrealized goal because the two most fundamental sets of physical principles, those of quantum mechanics and general relativity, are incompatible. The principles of quantum mechanics presume a flat space-time that is denied by general relativity. But even if someone succeeds in producing unified principles of quantum gravity, it is doubtful the resulting models could provide an exact fit to any significant portion of reality.

As I see it, the only way any particular model could exhibit an exact fit to the world is if it were a complete model that fits the world exactly in every respect. To see this, suppose we have a model that is not complete. That means that there are some things in the world not represented in the model. These unrepresented things may be expected to have some (perhaps remote) causal connections with things that *are* represented. But since these interactions are not represented in the model, the model could not be expected to be exactly correct about the things it does represent. So only a complete model could be expected to fit the world exactly.[16] But the prospects of ever constructing complete models of anything are remote. Even in the case of fundamental physics, it is not to be expected that the general principles of such a theory could specify such things as the exact distribution of matter in the universe or the relative prevalence of elements, for example, the ratio of hydrogen to everything else. For other sciences, even chemistry and biology, it is clear that their models capture only limited aspects of the world, leaving many unknown interactions to prevent any significant model from being exactly correct.[17]

The above statement of the argument invokes what seems like a meta-physical assumption of connectedness in the universe. The argument can be made less metaphysical by asserting only that we do not *know* the extent of connectedness in the universe. Nor could we know that a model is complete, that it has left nothing out. Indeed, even if we had a perfect model, experiment could not reveal this because every experiment introduces its own margin of error.[18]

I conclude that similarity remains an appropriate concept to employ when speaking of the general representational relationship between the world and models used to represent it. The concept of truth is not sufficient.

Generalizations

Some statements called "laws of nature" function more like lower level generalizations than grand principles. These also abound in physics; Hook's law, Snell's law, and Galileo's law of the pendulum being traditional examples. The prevalence of such laws goes down as one moves up Comte's hierarchy through chemistry, biology, and psychology to the social sciences. Even in physics, however, the simple statements of such laws cannot be both universal and true. There are always known restrictions and exceptions. The question is what to make of this situation.

One solution is not to take these law statements at face value, but to regard them either as being tacitly supplemented with imbedded *ceteris paribus* clauses or as being accompanied by separate qualifications.[19] A problem with this sort of solution, from my point of view, is that it requires being definite about something that is decidedly indefinite, and so the resulting package ends up being incomplete. Alternatively, in trying to be indefinite, this approach risks making laws vacuous, claiming, in effect, that the law holds except where it does not.

A better solution, I think, is to keep the simple law statements but understand them as part of the characterization of an abstract (representational) model and thus being strictly true of the model. The required qualifications, then, concern only the range of application of the model. One need only indicate, tacitly or explicitly, where it applies or not and to what degree of exactness. One might wish to claim, for example, that a whole class of previously identified mass-spring systems can be represented using the same type of model. And this could be tested directly by measuring periods of randomly selected members of the class. Of course, we now know enough about such systems that direct tests are seldom necessary.

In sum, in place of lower level laws of nature interpreted (mistakenly) as universally true generalizations, my account has the usual statements of such laws being exactly true of abstract (representational) models and then adds families of restricted generalizations regarding systems, or classes of systems, that the models fit more or less well.

Models of Data

As emphasized by Pat Suppes (1962) many years ago and James Woodward (1989) more recently, models are typically compared not directly with experimental data, but with *models* of data. For example, our model might include a number of variables, two of which are related linearly by the equation $y = ax$. A real system to which we wish to apply the model might then exhibit two measurable quantities, x_o and y_o, which we tentatively identify with x and y, respectively. A set of measurements yields pairs of data points of the form (x_o, y_o). These points might be graphed. A standard statistical software package, using a least-squares algorithm, may be used to compute the "best" estimate of the observed slope, a_o. This yields a linear model of the data. This model of the data incorporates the assumption that differences between real values and measured values are normally distributed.

The question now is whether the observed difference, d_o, between a and a_o is small enough to say that the measured value agrees with the value expected if the model indeed provides a good fit to the real system. Another statistical package could tell us how likely it would be to obtain the observed value, d_o, if in fact the real difference were zero. This requires some estimate of the statistical variance in the measurement process. Providing the required assumptions, we might calculate that the probability is 95 percent that we would obtain a value of d_o or less if in fact $d = 0$. On this basis, we could decide that the predicted value of a agrees well enough with the observed value to conclude that the model as a whole indeed fits the real system. This latter conclusion applies not only to the variables x and y, but to all other variables in the model. We conclude they all fit proportionately as well as x and y. That is what it means to say that the model as a whole fits the system of interest.[20]

The main point here, however, is that experimentally testing the fit of a model to some real system is a matter of comparing aspects of the model not with data directly, but with a model of the data. It is a model-model comparison, not directly a model-world comparison. And, of course, there may be several different legitimate ways of analyzing the data to obtain a model of the data. And different statistical techniques may be used in deciding when the observed agreement is sufficient to infer a general fit between the model and the real world. Different fields of inquiry may adopt different conventions about what are the proper ways of judging goodness of fit. These conventions may have a pragmatic rationale within that field. But there seems to be no one way of making these judgments that can be determined to be uniquely best across all areas of the sciences.

Theories and Laws

In my picture of scientific theories, there are no single elements explicitly designated as being "The Theory" or as being "A Law." This is because the terms *theory* and *law* are used quite broadly both in scientific practice and in meta-level discussions about the sciences. Their use typically fails to distinguish elements that I think should be distinguished if one is to have a sound metaunderstanding of scientific practice. Thus, for example, references to "evolutionary theory" may often be understood as referring to what I am calling "the principles of evolutionary theory." I regard these principles as defining a quite abstract object and not as directly referring to anything in the world. But just about everyone would insist that evolutionary theory is an empirical theory. In my terms, this means that some specific evolutionary models structured according to evolutionary principles have been successfully applied to real populations. So, from my point of view, the term *theory* is used not only ambiguously, but in contradictory ways. My account removes the contradictions.

The same holds for the term *law*. What is commonly called "Newton's second law of motion," for example, is for me a central principle of classical mechanics. The so-called law of the pendulum, on the other hand, is an explicit part of the characterization of a much more specific, though still abstract, representational model of the simple pendulum. So, here again, I prefer to leave the term *law* as it is and use a more precise vocabulary in my own account of scientific theorizing.[21]

Laws of Nature

The notion of a "law of nature" figures prominently in many accounts of the nature of science. Recall, for example, that Weinberg invoked laws of nature as statements expressing the known truths of physics. Thus, although I have already said what roles statements called "laws of nature" may play in my account, something more needs to be said.[22]

It is first necessary to be clear that the notion of a law of nature is not part of the literal content of any science. It is a notion that belongs to a meta-level interpretation of what scientists do, for example, that they discover laws of nature. Thus, to question the applicability of this notion is not to question any science itself but, rather, only how the aims and achievements of scientific activities are to be described. At this level, the opinions of scientists are to be considered, but they are not definitive. Here others may better describe what scientists do than scientists themselves.

One thing seems pretty clear. The notion of a law of nature did not arise out of the practice of science itself. Sometime in the seventeenth century, it was imported into discourse about science from Christian theology, both directly, and indirectly through mathematics. Originally, laws of nature were understood as God's laws for nature. Thus, behind the laws of nature was a lawgiver, from whence came the universality and necessity of such laws. It is still common for scientists and commentators on science to speak of nature as being "governed" by laws of nature, which suggests that somewhere there might be a "governor." It also seems pretty clear that this theological notion of laws of nature became prominent through the writings of Descartes. Newton then picked up the idea from Descartes, though he hardly needed Descartes to encourage his theological inclinations. How the notion got stripped of its theological implications throughout the eighteenth and nineteenth centuries seems a story yet to be told. Twentieth-century philosophers struggled to understand the nature of natural laws, especially features such as universality and necessity, in completely secular terms (Weinert 1995).

On its way to becoming secularized, the notion of a law of nature seems also to have acquired an honorific role. This is particularly clear in cases where the law incorporates the name of its putative discoverer (Snell's law, Ohm's law, etc.). This practice helps explain how statements that play rather different roles in science all got to be called "laws of nature." Rather than struggling to give a unified account of the resulting heterogeneous set of statements called laws of nature, my account separates out these roles. To summarize, I regard some statements called laws as principles that define highly abstract models. Newton's laws of motion provide a canonical example. Other statements define more specific models that embody a set of principles. Here the laws defining a simple harmonic oscillator provide a clear example. There are lots of empirical generalizations over kinds of things that approximate simple harmonic motion. Still other statements called laws define models without embodying any general principles. Growth curves for populations in ecology might be a good example of this sort of law. Finally, there are limited empirical generalizations embodying no known principles. Much of our medical knowledge, that aspirin cures headaches or that smoking causes lung cancer, is of this sort. I would not claim that these examples are exhaustive of things that might be called laws, nor even that they are exclusive. Before Newton, Galileo's law of the pendulum defined a simple mechanical model without being an instance of any known principles.

Fitness

I would like to pause briefly to consider the biological notion of fitness. And for two reasons. First, it provides a good illustration of the various roles principles and representational models play in scientific theorizing. Second, it provides a good analogy for understanding the fit of models to the world.

The principles of evolutionary theory can be compactly set out as follows:

A population, P, of organisms will evolve by natural selection if:
1. There are variations in phenotypes among members of P.
2. Some variations in phenotypes confer a reproductive advantage (relative to a shared environment) on those individuals possessing those phenotypes.
3. Variations in phenotype are to some degree transmitted to offspring.

The second condition is often understood as stating that, because of differences in phenotype, some individuals are *fitter* than others (relative to a shared environment). In short, there is differential "survival of the fittest." It has long been debated whether the survival of the fittest is an empirical law or a tautology. By calling these statements "principles," I am saying that they are not empirical laws, but that they together *define* what it is for a population to evolve by natural selection. They define a highly abstract model of natural selection. They do not by themselves say anything about any population of real organisms.

To formulate empirically testable hypotheses, one must designate particular populations and specify which properties of members make them fitter in their common environment. Darwin's finches provide a canonical example (Lack 1945; Grant 1986; Weiner 1995). The populations are those of several species of finches on the Galapagos Islands. The traits in question are the size and shape of beaks. The environmental features include the available food sources, such as plant seeds. With these specifications, we now have evolutionary models for the evolution of various species of finch. The models tell us, for example, that the strength and shape of beaks in various species should correspond to the relative size and hardness of the available seeds—which is indeed what is found among the real populations.

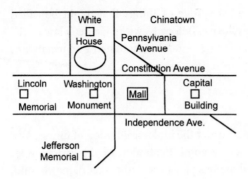

Figure 4.2 A partial tourist map of central Washington, D.C.

The parallel with the "fit" of models to aspects of the real world is this. To ask whether a model fits its intended object is to ask an abstract question. The question becomes answerable only once the specific features of the model being compared with something in the world are identified. In addition, some measure of closeness of fit must be assumed. But this measure can be specified only relative to the feature being examined. In the case of the finches, for example, a rough positive correlation between strength of beak and hardness of seeds is enough. The correlation need not even be expressed quantitatively. Model-based predictions of the period of a harmonic oscillator, on the other hand, can be evaluated quantitatively.

Maps

It helps in understanding the role of representational models to think about maps. Maps have many of the features of representational models. Figure 4.2 shows part of a standard tourist map of central Washington, D.C.

Consider some relevant properties of maps. First, maps are not linguistic entities. They are *physical objects*, for example, a piece of paper with colored lines and spaces on it. It does not, therefore, make sense to ask whether a map is *true* or *false*. Those designations (in English at least) are usually reserved for linguistic entities. Moreover, maps are not usually thought of as *instantiations* of any linguistic forms. They are not models in the logicians' sense.

Nevertheless, even though they are neither linguistic entities nor instantiations of linguistic entities, maps are *representational* in the sense that they can be used to represent a geographical area—as anyone who has used a map when traveling in unfamiliar territory knows full well. Just *how* maps manage to be used representationally is another question, one I will take up shortly. First, consider some further characteristics of maps.

Maps are *partial*. Only some features of the territory in question are represented. For example, the map of Washington, D.C., in figure 4.2 represents very few of the many buildings in the capital. Moreover, even the indicated features are not fully specified, such as the height of the Washington Monument.

Maps are of *limited accuracy* regarding included features. Relative distances on the map, for example, will not correspond exactly to relative distances on the ground. This could not be otherwise. No real map could possibly indicate literally all features of a territory with perfect accuracy. In the limit, the only perfect map of a territory would be the territory itself, which would no longer be a map at all.[23]

Let us now return to the question, *How* is it that the map of figure 4.2 can be used to represent Washington, D.C.? Part of the answer is: by being *spatially similar to* aspects of Washington, D.C. For example, the map shows the Washington Monument as being between the Lincoln Memorial and the Capital Building. And in fact this is so. In using a map we are using features of one two-dimensional surface (the map) to represent features of another two-dimensional surface (the surface of the city). For another example, some lines on the map represent streets in the city. Pennsylvania Avenue, for example, is represented as intersecting Constitution Avenue. And it does. To generalize: A map represents the region mapped partly in virtue of shared spatial similarities between the map and the region mapped. Here one *object* (a map) is used to represent another *object* (a geographic region). This notion is explicitly opposed to that of a statement representing a state of affairs by being *true of* that state of affairs. This holds in spite of the fact that, if it is indeed similar in the stated respects, the map incorporates the truth of many statements, for example, the statement that Pennsylvania Avenue intersects Constitution Avenue.

Spatial similarities are not intrinsic to maps and geographical regions. The *respects* in which similarity is claimed must be specified by the mapmakers or map users. Moreover, for any particular respect, there must be at least some implicit standard for degree of similarity, or accuracy. Standard road maps, for example, are very good at giving relative *distances*, but generally provide only rough indications of *elevation* and often no indications at all. For most automobile drivers, distances are more important than elevations. This is not so for bicyclists, for whom relative elevations are very important. In general, the features of the terrain that are mapped and the accuracy with which they are mapped are *interest relative*. There is no such thing as the objectively correct map of any place apart from the interests of intended users.

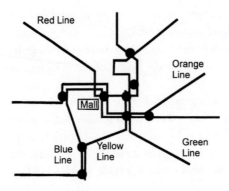

Figure 4.3 A partial subway map of Washington, D.C.

This latter point is evident in the differences between the tourist map of Washington, D.C., and the partial subway map shown in figure 4.3. There is, of course, a rough correspondence between the two maps, but the main information carried by the subway map is not distance but the topological ordering of the stations.[24] When boarding, one wants to be sure that one's destination is among the stops for this particular train. Then one wants to know when one is at the stop before one's desired destination. For this, one has to listen for the announcements or look out the window for the names of previous stations.[25]

Mapmaking and map using also depend on *conventions for interpretation.* We in contemporary Western culture are so used to using maps that we are hardly aware of the amount of convention involved in using maps. Here it helps to consider maps from other times and other cultures. Figure 4.4 shows a schematic version of a type of map that existed in various forms in Medieval Europe from roughly 500 to 1500 (Woodward 1987).

If one did not know this was a map, one might think it just an emblem of some sort. In fact, it was intended to be a map of the world (originally called *mappaemundi*), now called a T-O map. The outer circle represents the world's oceans. The top part inside the outer circle represents Asia. The bottom left represents Europe and the bottom right, Africa. The area between the two bottom regions represents the Mediterranean Sea. So east is represented as being at the top of the map, and north is to the left. Between the representations of Europe and Asia is a representation of the River Don; between Africa and Asia is the River Nile.

Note that I just wrote "Africa," "Asia," and "the River Nile" rather than "representation of Africa," and so forth. I would wager no reader was misled. Our mapping conventions are often quite transparent to us. And part of the reason they are so transparent is that they are part of our cultural heritage. It is no accident, for example, that in the United States and Western

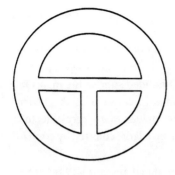

Figure 4.4 A medieval T-O Map.

Europe, world maps are centered on the Atlantic Ocean. What maps represent, and how, is culture dependent.[26]

Not only do we need conventions for interpretation, we need to know what a map is supposed to be a map of. The connection between a map and the territory mapped requires an intentional designation. Consider the simple line drawing of figure 4.5. I claim it is a map of the United States. The small internal rectangle represents my own state of Minnesota. I could use this map to show a European friend roughly where Minnesota is relative to the rest of the United States. The shapes of New England, Florida, California, and Texas are irrelevant to this question. So what makes figure 4.5 a map of the United States rather than just a line drawing? Partly, the rough similarity between its shape and that of the United States, but also partly my *intention* to use it as such.

Now I would like to say that the cultural background, the conventions for mapmaking, the designation of the region mapped, the specification of what features are mapped, and the degree of accuracy all determine a *perspective* from which the region is mapped. Every map reflects a perspective on the region mapped, a perspective built in by the mapmakers. In short, mapmaking and map using are *perspectival*.

Lessons from Cartography

Many of the features of maps just examined are explicitly recognized in the practice of cartography. Here I will note only a few.[27] The first is generally

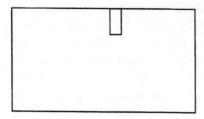

Figure 4.5 A line drawing used to represent parts of North America.

called "feature selection." Before making a map, the mapmakers must decide just what kinds of features of a terrain they intend to represent. Not all known kinds of features can be represented, and often not even all features of a kind that are represented. How is it decided what features to represent? Here, a strong constraint is provided by the intended *scale* of the map. On a large-scale map representing a relatively small area, such as the Federal Triangle in Washington, D.C., it is possible to include many special features specific to that area. A small-scale map covering a large country, by contrast, is restricted to many fewer features.

Once the constraints of scale are accommodated, the primary reason for including some features and not others is, in most cases, the *purpose* for which the map is being made. What are the intended uses of this map? Once it is decided what features are to be represented, there are further decisions regarding *how* to represent the desired features. Many of these latter decisions, however, are determined primarily by long-established conventions for different kinds of maps.

Another notable feature of cartographic practice is called "smoothing." Here again, the scale of the map is highly relevant. A small-scale map can, for example, represent numerous turns and twists in secondary roads. On a larger-scale map, these irregularities will be "smoothed out" so only the general directions of the roads are indicated. The same goes for rivers and shorelines.

Given the necessary feature selection and smoothing, maps end up being something that can be thought of as *models* of the areas they represent. This is dramatically evident in the case of three-dimensional relief maps of the sort sold at visitors' centers in American national parks. The analogy between maps and models goes both ways.

Maps and Models

I hope it is clear that scientific models share many important features of maps. Models, too, are objects, not linguistic entities such as sentences. Typical scientific models, however, are abstract rather than physical entities. Representational models also are not instantiations of linguistic forms. And they are not properly called true or false. But they can be used to represent things in the world, which requires determining to what sorts of things any particular type of model is intended to be applied. For scientific models, it is also similarity with some real thing that is intended, although, of course, not a simple similarity of spatial features. For models, the specific respects in which similarity is claimed must be specified, as must the desired degree of similarity. Again, for any particular type of model, there are implicit conventions for interpretation. Finally, the use of models is interest relative: what models are used depends on the purposes for which a representation is required.

The connection between maps and models in the sciences is often quite explicit. Indeed, in some fields, the products of scientific research are literally maps. In other fields the connection to mapping is somewhat metaphorical, although there is no doubt about the appropriateness of the metaphor. It is instructive to note the range of these examples. I will proceed from the literal to the more metaphorical.

Apart from geography itself, sciences that produce literal maps include geology, geophysics, and oceanography. This list should also include the planetary sciences, where scientists are busy mapping the surface of the Moon and planets, particularly Mars and Venus. The project of mapping the ocean floor, which was pursued vigorously after World War II, contributed greatly to the 1960s plate tectonic revolution in geology.[28] Geophysical "maps" of structures well below the surface of both the continents and oceans exhibit the overlap between traditional maps and scientific models. These maps are produced using a combination of theoretical models and data from various types of seismic instrumentation. Here one cannot say where mapping leaves off and modeling begins.

There were maps of the heavens long before the Scientific Revolution. Systematic mapping of the stars has been pursued since the invention of the telescope.[29] Remapping occurred with major developments in observational technology, including, in the twentieth century, the invention of radio telescopes. A recent entry in this progression is the National Aeronautics and Space Administration (NASA) MAP Project, where "MAP" stands for "Microwave Anisotropy Probe." From an orbit centered four times the distance from the Earth to the Moon, this probe measures microwaves coming from the whole sky, providing what NASA reported as "the first detailed full-sky map of the oldest light in the universe," revealing the universe as it was about 380,000 years after the Big Bang.[30] Interestingly, the resulting image serves both as a model of the data and as a model of the young universe. The difference is wholly a matter of interpretation.

It is no accident that neuroimaging is often referred to as "brain mapping" (Toga and Mazziotta 1996). This metaphor is particularly apt in the case of *structural* imaging done with techniques such as computer assisted tomography (CAT) and magnetic resonance imaging (MRI). After all, the images represent the relative spatial locations and shapes of physical structures that make up the brain. These images also provide indications of differences in the physical makeup of various structures, as indicated by their relative absorption of X-rays or relative ability to disrupt induced resonances in selected atoms. The notion of mapping is somewhat more metaphorical when the imaging is *functional* rather than structural, as with positron emission tomography or functional MRI. These instruments detect concentrations of agents introduced into the bloodstream or changes in the oxygen content of the blood itself, from which the location of various functions is inferred. Here again, the same image may be understood as a model of the data or interpreted as a spatial model of brain functioning.

Brain imaging research also provides an interesting example of the move from models of data to models of the object. There is considerable variability in the size and structure of individual human brains. Yet, for scientific purposes, it is convenient to have a standardized model of "the brain," or maybe "the male brain" and "the female brain." There is no uniquely best way to construct a standardized model from individual images.

Flattening the Earth

The need to "flatten the Earth" (Snyder 1993) in order to produce a two-dimensional, printable, map of the three-dimensional surface of the earth provides an instructive example of perspectivism. Conceived of as the problem of projecting all points on the surface of a sphere onto a flat two-dimensional surface, the problem has no uniquely best solution.[31] One can, for example, make the projection so that equal *distances* along the spherical three-dimensional surface are mapped onto equal distances along the flat two-dimensional surface. In that case, however, equal *areas* on the three-dimensional surface will not map onto equal areas on the two-dimensional surface. Other projections, however, will map equal areas onto equal areas, but at the cost of not preserving distances. It is a theorem of projective geometry that one cannot do both. Similar conflicts arise for other variables, such as scale and shape.

The most widely known projection is still that produced by the Flemish polymath, Gerardus Mercator in 1569 (Snyder 1993, 43–49), shown at the top in figure 4.6. For a great many people all over the world who have any familiarity with maps at all, Mercator's map is the one they know best. This is unfortunate because it presents a quite distorted picture of the geography of the Earth. But that is hardly Mercator's fault. He did not design his map for general geographical purposes. He designed it for use in navigation. In fact, the English translation of the map's original Latin title is "A New and Enlarged Description of the Earth with Corrections for Use in Navigation." Its virtue for navigational purposes is that, to sail from point A to point B, one can simply draw a line on the map from A to B, and then just follow the

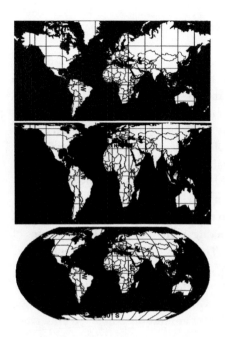

Figure 4.6 Three projections of
the Earth's surface onto two
dimensions: Mercator (top), Peters
(middle), and Robinson (bottom).
(Images courtesy of Peter H. Dana.)

indicated compass setting, making allowances for the difference between true north and magnetic north.

The distortions in area on the Mercator map are exemplified by the so-called Greenland Problem. On the map, Greenland appears to be about the same size as Africa. In fact, at only 0.8 million square miles, Greenland is less than one-fourteenth the size of Africa, measuring 11.6 million square miles. Less dramatic, but still disturbing, Europe appears to have about the same area as South America, when it is in fact only about four-sevenths as large. In 1973 a German historian and journalist, Arno Peters, published what he called a "new" map projection (middle map in fig. 4.6) that roughly preserved equal areas. Peters, and others, have promoted this map as eliminating the "Eurocentric bias" in the Mercator map, because it more accurately represents the relative land masses in both the northern and southern hemispheres, which it does. It is not completely implausible that the use of Mercator's map for general geographical purposes for over four hundred years was not simply due to its value to navigation (essential to European imperialism!), but also to an unconscious Eurocentrism. But Peters's map is neither new nor particularly good from a general cartographic perspective. A similar equal area projection was published already in 1855 by the English clergyman James Gall. Moreover, as is obvious, the Peters map achieves equal areas by introducing pronounced distortions in shape, relative to shapes on a globe.

In 1988 the National Geographic Society endorsed the Robinson map, shown at the bottom of figure 4.6. This map is not a simple mathematical projection at all, which is to say, there is no single formula that projects points on a sphere into points on a surrounding cylinder. Rather, it was adjusted so as to achieve a compromise among competing variables, particularly area and shape. Thus, while not strictly an equal-area projection, it provides about as good a general view of the relationships among land masses and oceans as one can achieve in any two-dimensional projection.[32] The major feature of this map, as opposed to both the Mercator and Peters maps, is that it gives up representing both lines of latitude and longitude as straight lines intersecting at right angles. Only the lines of latitude are straight and parallel. The lines of longitude, except for the central (Greenwich) meridian, are all curved, giving the map a globelike appearance in spite of being only two-dimensional. One might still argue that taking the central meridian at Greenwich exhibits a North Atlantic bias. It could in principle be anywhere around the whole circumference of the Earth.

Taking the problem of mapping the surface of the Earth onto a flat surface as an analog for the general practice of constructing models in science has both advantages and disadvantages. One advantage is that it provides a clear example of the idea that representation is representation *for a purpose*. Navigation and having a general knowledge of world geography are two very different purposes. There are different projections that serve each of these two purposes. None do a very satisfactory job of satisfying both purposes. For a general understanding of the gross geography of the Earth's surface, nothing beats a globe. But for navigation, a Mercator-style map is still more useful.

In general, the projection problem provides a good example of a kind of perspectivism. Every projection gives a different perspective on the Earth's surface. But these projections are all incompatible. They cannot, for example, simultaneously preserve shapes and areas everywhere. So they must differ somewhere in how they trade off these two variables. This feature of the example conflicts with what I presume to be the widespread methodological presumption among scientists that different perspectives on a single universe should, in principle, be compatible. This conflict seems to me due to the highly constrained nature of the example, that of projecting the surface of a three-dimensional sphere onto two dimensions. Scientific theorizing is not, in general, so highly constrained—or so we presume.

Truth within a Perspective

There is an argument that has often been used to discredit perspectivism (Hales and Welshon 2000, chap. 1). Is perspectivism itself true or false? If it is true, then there is at least one claim that is nonperspectivally true. So if perspectivism is true, then it is false. Of course, if it is false, then it is false. So perspectivism is false.

As I see it, this argument simply begs the question. It assumes that truth is to be understood in objectivist terms, which is just what perspectivism denies. For a perspectivist, truth claims are always relative to a perspective. This is not so radical a view as it might sound. It was long a doctrine within Logical Empiricism, and analytic philosophy generally, that scientific claims are always relative to a language. First one chooses a language, then one makes claims that may be judged true or false. And the choice of a language is pragmatic, not itself a matter of truth or falsity.

My claims on behalf of perspectivism are not much different. Only rather than focusing on language, I focus on the physical characteristics of instruments (including the human visual system) and principles defining generalized models. Of course one uses language to talk about observations and theoretical claims, but this is a mishmash of everyday language and scientific terms. It is not the language that determines the perspective. And the notion of truth is used only in a minimal, and decidedly not metaphysical, fashion.

Consider an example using the notion of perspective in a relatively literal sense. Imagine standing on the steps of the U.S. Capital Building looking down the Mall toward the Washington Monument. From that vantage point, the White House is off to the right of the Washington Monument. Is that true? It surely is not false. Just inspect the map of central Washington, D.C., sketched in figure 4.2. Anyone who denies that the White House would be off to the right would be just plain wrong. As noted several times earlier, what more one should say about this use of the notion of truth is a controversial philosophical matter (Künne 2003). What is clear is that we are not forced to adopt an objectivist metaphysics in order to declare false the claim that the White House would be to the left of the Washington Monument.

In this simple example, relativization to a perspective is quite natural. We need only add something like "From where we stand . . ." That is easily understandable. And so is the fact that things would be reversed if one were standing on the steps of the Lincoln Memorial looking toward the Washington Monument. In that case the White House would be to the left. Of course, we do not typically think of our everyday ways of speaking as being within a perspective. But neither do most people think of colors as being relative to the perspective of the human chromatic visual system. That scientific observational or theoretical claims should in general be relativized to a perspective is, if anything, easier to accept.

Perspectives and Paradigms

Anyone at all familiar with developments in the history, philosophy, or sociology of science during the second half of the twentieth century is likely to have wondered what connections there are between what I am calling perspectives and Thomas Kuhn's (1962) paradigms. There are some important similarities. Claims about the truth of scientific statements or the fit of models to the world are made within paradigms or perspectives. Here, a comparison with data or an observational perspective can be decisive. On the other hand, claims about the value of paradigms or perspectives themselves are not subject to anything like decisive observational tests. At most they become dysfunctional, failing to generate new verified findings or models fitting new phenomena. Eventually they may be displaced by new paradigms or perspectives that provide resources for new research leading to new verified claims or successful models. These are the major similarities. Now for the differences.

Kuhn's use of the term *paradigm* was notoriously ambiguous.[33] In a postscript to the second edition of *The Structure of Scientific Revolutions*, Kuhn himself distinguished two very different senses of the term. "On the one hand, it stands for the entire constellation of beliefs, values, techniques, and so on shared by the members of a given [scientific] community. On the other, it denotes one sort of element in that constellation, the concrete puzzle-solutions which, employed as models or examples, can replace explicit rules as a basis for the solution of the remaining puzzles of normal science" (1970, 237). The first of these elements Kuhn called a "disciplinary matrix"; the second, "exemplars." Neither of these two senses corresponds with either observational or theoretical perspectives as I understand them. A disciplinary matrix is much broader than a perspective; an exemplar is much narrower, corresponding more closely with the application of a particular type of representational model.

Incommensurability

Among Kuhn's most contentious claims, one that continues to generate controversy was that paradigms (now understood as disciplinary matrices) are "incommensurable."[34] There are two generally accepted interpretations of this claim, one linguistic, one methodological. The linguistic interpretation is that the *meanings* of terms in different paradigms are different and not intertranslatable. The term *mass* in both classical and relativistic mechanics provides a standard example. The methodological interpretation is that the *standards* for judging the truth or acceptability of claims are irreconcilably different in two different paradigms. Here I will consider only the linguistic interpretation.

My view is that the whole idea that there might actually be linguistic incommensurability in the practice of science itself is an artifact of the historical circumstances in which historians and philosophers of science operated in the second half of the twentieth century. When Kuhn himself, as a physicist turned historian, looked at physics before the seventeenth century scientific revolution, he was struck by the difficulty of understanding what they were talking about. Nevertheless, in his book on the Copernican revolution (1957), he did a very good job of explaining Ptolemaic astronomy as embedded in an Aristotelian worldview. Why would he think that Galileo had any difficulty understanding Ptolemaic astronomy or Aristotelian physics?

The answer to this latter question seems to lie in the philosophies of language to which Kuhn was exposed at Harvard and then later at Berkeley. Two features of these philosophies are especially relevant. First was the doctrine that the meaning of terms is *holistic*, that is, the meaning of any term is a function of its place in a network of terms connected analytically or, for many who, following Quine, rejected an analytic-synthetic distinction, by inferential relationships of all kinds, including inductive relationships. Second was the doctrine that understanding another language requires being able to *translate* that language into one's own. Kuhn and others doubted, for example, that it is possible to translate the classical term *mass* into the language of special relativity, in part because of the obvious differences in inferential connections with other terms.[35] I am not the only student of these developments to conclude that approaching the study of science through the philosophy of language was counterproductive.[36]

It is clear, I think, that there are no problems of linguistic incommensurability for perspectives. There are genuine problems in comparing the outputs of different instruments, for example, PET and MRI scans, but these are scientific problems for which there are various recognized solutions. And theoretical scientists have no particular linguistic problems in switching from classical mechanics to relativistic mechanics, from classical electrodynamics to quantum mechanics, or from classical genetics to molecular genetics. Comparing perspectives and switching from one to another are a normal part of scientific practice. We don't need a *theory* of language to recognize this practice, but it would be nice if there were such to deepen our understanding of it.[37]

The beginnings of an appropriate theory of language might be found not in the philosophy of language but in cognitive linguistics (Lakoff 1987) and so-called usage-based approaches to language (Tomasello 1999, 2003). In such theories, the emphasis is on the use of language for communication in social interactions. There is a particular emphasis on language acquisition. This suggests a computer-like analogy.

A human is, among other things, a cognitive system. In learning an everyday language, a human acquires a kind of operating system for engaging in other activities. One of these activities might be designing and using instruments for scientific observation. Another might be employing scientific principles to construct models of various natural systems. We might think of learning to engage in such activities as involving the acquisition of various applications programs. Now the payoff of this way of thinking is that we realize that learning to use an application program does not require translating that program into the language of the operating system. Nor does using different application programs require translating one into the other. It is just a matter of acquiring the skill of using the various programs, and this initially requires only knowing some everyday language. We can forget about linguistic incommensurability as a special problem for understanding the activity of doing science.

Scientific Kinds

Differences between objectivist, constructivist, and perspectival understandings of scientific claims are highlighted by discussions of what, since the mid-nineteenth century, have been called "natural kinds."[38] Here I will ignore constructivist approaches and concentrate on the difference between objectivist and perspectival approaches.[39]

Objective Natural Kinds

It seems to be implicit in an objectivist account of scientific claims that the goal of scientific research is to develop concepts that correspond with objective natural kinds, kinds of things that constitute objective nature apart from any human attempts to describe it. Presumably, the set of objective natural kinds would be unique. I agree with Hacking (1991), who cites Strawson (1959, 169), that it is difficult to make sense of this notion of natural kinds. It seems to require that there are uniquely correct terms or descriptions that pick out objectively real natural kinds. It is as if science were in the business of discovering the language God used when he named the beasts of the field in the Garden of Eden.

Two areas of science have provided the bulk of the examples used in philosophical discussions of natural kinds, biology and chemistry. In biology the focus has been on the nature of species.[40] There now seems to be a near consensus among both biologists and philosophers of biology that the classical approach to determining kinds, specifying a set of intrinsic properties each necessary and all jointly sufficient, does not work for species. Members of paradigmatic species are just too variable—as indeed they should be if evolution is to work. The closest one can come to an essential property for a species is a *relational* property, being a member of a historical population of interbreeding individuals. This way of thinking invites us to drop "species" as a fundamental concept in biology and simply focus on evolving populations. From this perspective, a primary mechanism for what is commonly called "speciation" is simply the, usually accidental, physical isolation of a small subgroup from the main population. From that point on, the subgroup and the main population evolve independently until members of the two populations are no longer capable of interbreeding. If one thinks in terms of species, it seems arbitrary whether we say that the original species, having split into two new species, has ceased to exist or that the original species continues to exist while the offshoot evolves into a new species. Does the situation involve two or three species? The prospect of settling on a unique set of objective natural kinds seems remote. More fundamentally, the basis for thinking in terms of evolving populations is evolutionary theory. Therefore, consideration of what kinds do or do not exist seems to be based on theory. Thus, in the case of biology, thinking about kinds does not further the argument for an objectivist understanding of scientific claims. The reasons for thinking that theorizing is perspectival still stand.

The situation is similar for chemistry. LaPorte (2004, chap. 4) has argued persuasively that chemical *compounds*, such as water and jade, are not good candidates for objectively natural kinds. Chemical *elements* such as hydrogen, oxygen, and gold are often thought to be better candidates. Better candidates, maybe; but not sufficiently good candidates. The periodic table of the elements classifies elements by their *chemical* properties. How do they react chemically with other substances? It turns out, of course, that many chemical properties are linked with atomic *number*, because chemical reactions are a function of interactions among electrons (chemical bonding) and the number of electrons in an element equals the number of protons, which defines atomic number.

But there are some chemical properties, and many physical properties, that are more a function of atomic *weight* than atomic number. These include rates of chemical reaction, the melting point of solids, the boiling and freezing point of liquids, and rates of diffusion in solution. Atomic weight is the sum of the number of protons and neutrons in a nucleus. Elements with the same atomic number but differing atomic weights (same number of protons, different numbers of neutrons) are called *isotopes* of the respective elements. There is no recognized name for isotopes with the same number of neutrons, presumably because there are few interesting properties that correlate with what we might call "neutron number." The operative term here is "interesting," meaning interesting to humans, because the determination of objectively natural kinds is supposed to be independent of any human interests. Yet there seems no way of deciding whether the objective natural kinds are those determined by atomic number, atomic weight, or even neutron number. Nor is this a matter of lack of knowledge. There does not seem to be anything that we might find out about the elements that would determine which are the objective natural kinds.

Faced with these difficulties, it is tempting to follow Churchland (1989, chap. 13) in suggesting that maybe the only really objective natural kinds are elementary particles. But this won't do either, because what counts as an elementary particle is highly dependent on current fundamental theory. That takes us back to the basic problem of whether claims based on theories are objective or perspectival. So, once again, focusing on natural kinds fails to provide support for an objectivist understanding of scientific claims.

Theoretical Kinds

The assumption that natural kinds are more basic than theories follows from a simple empiricist understanding of theories as organizing natural laws, which in turn are generalizations over existing kinds. So, first kinds, then laws, then theories. This is how "All ravens are black" became a philosophical exemplar of a scientific law of nature. More recent philosophical commentators on natural kinds have abandoned this simple picture and tend to emphasize the role of theory (or causality) in the determination of what counts as a natural kind (Bird 1998, chap. 3; Boyd 1991; Griffiths 1997, chap. 7; LaPorte 2004; Psillos 1999, chap. 12). These authors vary in the extent to which they are concerned to advance an objectivist, or at least a metaphysical, understanding of natural kinds. Yet they all retain the assumption that theories consist of descriptive statements to which the concept of truth is unproblematically applicable. A model-based understanding of theories provides grounds for a different understanding of theoretical kinds.

Abstract Models: Defined by the Principles of Classical Mechanics (Newton's Laws).

Conservative Models: Defined by the constraint that the Total Energy of the system remain constant.

Uniform Rectilinear Accelerating Motion.	Harmonic Motion.	Orbital Motion.
Defined by the constraint $F = k$.	Defined by the constraint $F = -kx$.	Defined by the constraint $F = -k/r^2$.

Figure 4.7 Mechanical kinds generated by the principles of classical mechanics.

The clearest case is that in which a theory is based on a set of general principles defining a very general set of abstract models. Classical mechanics again provides an exemplar. Beginning with the general principles, basically Newton's laws, one can define an open-ended set of more or less general models, in fact, nested sets of models. Figure 4.7 describes a selection of such models. The general model for a simple harmonic oscillator, defined by the force function, $F = -kx$, for example, yields more specific models for a simple pendulum, a bouncing spring, a vibrating string, and for many more oscillating systems. Typically, these models do not follow deductively from the basic principles of classical mechanics. Approximations (simplifications, idealizations, etc.) are often involved; for example, for small angles the sine function of an angle may be replaced by the angle itself. Nevertheless, the resulting models are well-defined abstract entities. Galileo's law of the pendulum holds exactly for a model of a simple pendulum.

My view is that each of these models represents a *kind* of mechanical system. The equations of motion for these models define kinds of mechanical motion. We therefore have a clear notion of theoretical kinds. They are defined using the principles of the relevant theory. So the theory and implied laws come first. The kinds are defined relative to the theory.[41]

Determining the empirical counterparts of theoretical kinds is another matter altogether. Since no real system ever exactly matches the ideal behavior of a theoretical model, we cannot simply say that the corresponding set of empirical objects of the defined kind are those whose behavior matches that of the model. What we count as being empirical members of the corresponding set of real systems depends on how good a match we require and in what respects. These judgments cannot help but be interest relative. The period of a mechanical oscillator in a piece of precision equipment might have to be within one part in a million of the ideal. In a child's toy, one part in fifty might be good enough. So the set of empirical harmonic oscillators cannot be sharply delimited. The good news is that it need not be so for general purposes. We can usually get by well enough with vaguely defined empirical kinds. If more precision is required for specialized purposes, we can often achieve it.[42]

Perspectival Realism

As noted in chapter 1, philosophers of science debated the merits of scientific realism throughout the second half of the twentieth century, with a number of book-length treatments appearing at the end of the century. Often, debates between philosophers of science and recent sociologists of science have been framed as a conflict between *realism* and constructivism.[43] As I indicated in chapter 1, I now think this is not the best way to frame the opposition. It is better to think of it as an opposition between what I have been calling *objectivism* (or objectivist realism) and constructivism.

Among philosophers of science, the problem has been primarily one of *epistemology*, namely, of how one might *justify* inferences from what can be observed to what can only be described theoretically. So philosophical anti-realism has been a species of traditional philosophical skepticism.[44] Few philosophers would today adopt a wholesale skepticism about the existence of an external world. Wholesale skepticism about so-called theoretical entities now seems to me almost as quixotic. A philosopher might raise questions about the reasonableness of accepting some of the more exotic claims of a few theoretical physicists, but these would have to be the kind of questions that other physicists might raise. When it comes to genetics or neuroscience, philosophical skepticism must surely give way to standard scientific practice, which is decidedly realistic (but not necessarily objectivist) about all sorts of "theoretical" entities.

There remains, however, another problem for philosophers of science and for students of the scientific enterprise more generally, namely, understanding how scientists in fact go about justifying their claims as to the relative fit of specific models to aspects of the world. Here the task is not to provide a *philosophical* justification of scientific methods but merely to *describe* how these methods work and to *explain* why they are thought to be generally effective. A good account of these methods, even at a fairly high level of abstraction, supports a perspectival rather than an objectivist understanding of scientific realism.

The Empirical Testing of Specific Models

Only *specific* models can be directly tested empirically. These tests provide the basis for the indirect empirical evaluation of more general or more abstract models. In the most general terms, I would describe the empirical testing of models as a process of bringing together two perspectives, one observational and one theoretical, in order to decide whether the model fits the world as well as desired. The initial presumption is that the observational perspective has priority. The models of data generated within the observational perspective are to be used to decide on the fit of the model generated by theoretical principles; not the other way around. But this is only a strong methodological presumption. The theoretical model might in some cases be used to question the reliability of the observational instrumentation. Here I will only consider the more usual case.

As indicated in chapter 3, principles of nuclear physics together with principles of astrophysics have been used to generate models of star formation. Among these models are some that include the generation of heavier elements such as Al 26 in very massive stars, such as those thought to exist at the center of the Milky Way. Al 26 is known, both theoretically and experimentally, to decay into Mg 26 by emitting a gamma ray with energy 1.8 MeV. The COMPTEL instrument was able to detect and record gamma rays of this energy and produced the image shown in plate 6, indicating a noticeable flux of 1.8 MeV gamma rays coming from the region around the center of the Milky Way. This result was taken as a basis for deciding that the indicated models of heavy element formation in large stars do pretty well fit the actual processes of heavy element formation.

Why Standard Experimental Tests Are Effective

It remains to provide an explanation of why standard tests of theoretical models are effective in determining whether or not to regard a particular model as

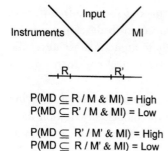

$P(MD \subseteq R \ / \ M \ \& \ MI) = High$
$P(MD \subseteq R' \ / \ M \ \& \ MI) = Low$

$P(MD \subseteq R' \ / \ M' \ \& \ MI) = High$
$P(MD \subseteq R \ / \ M' \ \& \ MI) = Low$

Figure 4.8 Schematic rendering of an experimental test of model fit.

indeed providing a good fit to its subject matter. My explanation continues to support the general position that the scientific conclusions are always relative to theoretical and instrumental perspectives. I think any plausible rival explanation of the effectiveness of experimental tests would end up supporting a perspectival rather than objectivist understanding of scientific claims.

Consider the situation shown schematically in figure 4.8. The funnel represents the instrumentation into which we are to imagine some part of the system under investigation to be inserted, in this case radiation from the center of the Milky Way. What comes out is data that are used to construct a model of the data, *MD*, which we will assume for simplicity constitutes a segment of a one-dimensional line. This would literally be the case if we were measuring a single parameter. *MD* is a line segment rather than a point, of course, since every measurement has some dispersion. We presume that, employing the relevant models of the subject matter, *M*, together with models of the instrumentation, *MI*, one can determine a region of the line, *R*, which includes the expected (most probable) location for *MD*, *if* our models of both the subject matter and the instrumentation do indeed fit well. Ideally, the two regions, *R* and *MD*, should be fairly small for a decisive test.

Philosophers of science have typically attempted to understand empirical tests in terms of logical relationships among statements describing things like *M* and *MD*. This is in accord with the attempt to understand representation as a two-place relationship between the same sorts of statements. Just as I prefer to understand representing as an activity involving agents using models to achieve various goals, so I prefer to understand testing models as an activity in which agents use experiments to *decide* whether or not to regard a model as providing a good fit to the objects of inquiry. This involves, among many things, deciding how probable results in the region *R* should be, and thus how wide *R* should be. This in turn requires trading off risks for two possible mistakes, namely, concluding there is a good fit when there is not and concluding there is not a good fit when there is. There are no general rules for making these trade-offs. That depends on the context. Here I am not concerned to develop a general account of such decision making, but merely to argue that, whatever the details, the decisions remain within given theoretical and instrumental perspectives.[45]

If *MD* turns out to be decidedly *outside* the region *R*, scientists are going to conclude that *something* is wrong. Either the models in question don't generally fit the world as well as they had hoped, or the instrumentation was not working as they think it should, or maybe they miscalculated what the expected data should be like, or maybe there was a mistake in the data analysis leading to the model of the data, or who knows what? A fault analysis is in order. Eventually they may indeed conclude that everything was working as expected and that they must take the observed result as counting *against* a decision to accept the claim that the models in question do fit the overall real situation, even approximately. This judgment is pretty much just common sense and does not require an objectivist metaphysics. Whether this leads scientists to doubt the overall viability of the perspective within which the models were constructed is another matter.

What if MD turns up in the region R, as was the case with the COMPTEL observations pictured in plate 6? Were the scientists in question taking this situation, by itself, as the basis for a decision to accept the claim that the solar models in question do pretty well fit the nuclear processes in stars at the center of the Milky Way? If so, there is a well-known philosophical problem with this all too easy move. Might there not be different models of the subject matter that also generate the expectation that MD should turn up in region R? If so, taking this result as favoring a decision to accept the original models could be a mistake.

I do not think that the scientists in question were making this simple mistake. Rather, given their knowledge of nuclear physics and star formation, they know that any plausible rival model developed within these frameworks would make it *unlikely* that there should be a high flux of 1.8 MeV gamma rays coming from the center of the Milky Way. Thus, the situation is more like that pictured in figure 4.8, with any plausible alternative model, M', yielding an expected MD in a region, R', clearly outside the region R.

The overall procedure just sketched has highly desirable features. If M does not fit the world very well, it is expected that the resulting MD will fall outside the region R, and scientists will conclude, rightly by their assumptions, that this counts against deciding that M fits the real situation. If M does fit the subject matter well, they expect MD to end up in region R, and they will conclude, correctly, that this counts in favor of deciding that M does fit. Similarly, if some M' provides a better overall fit to the subject matter, it is unlikely that MD will turn up in region R and likely that it will turn up in R' instead. Either way, given their assumptions about the experiment, they have a reliable basis for deciding whether the observations count in favor of or against the claim that M fits the relevant aspects of the world.[46]

The important point for my purposes here is that all of these judgments remain within the specified observational and theoretical perspectives.

Thus, the strongest claim that can justifiably be made has a general form something like: Given the assumed observational and theoretical perspectives, M exhibits a good fit to the subject matter of interest. There is no basis for going further to a supposed more objective, nonrelativized, claim that this is how the world is, period.

As outlined above, the alternatives to M were all constructible within existing nuclear and astrophysical perspectives. But the overall situation would not change much even if there were another perspective (or several other perspectives) within which to construct alternatives to M. The concluding decision would then be relativized to several rather than just one perspective. It would not be metaphysically objectivist.

I am now in a position to suggest that much of the philosophical literature on scientific realism suffers from the unanalyzed assumption that a robust scientific realism must be objectivist realism because otherwise it slides into constructivism or relativism. But in an absolute objectivist framework, the problem of alternative models is acute because the realm of logically possible alternatives is too rich. It cannot even be described analytically. Antirealists delight in this situation. Realists struggle to overcome it. The problem disappears, however, if one is willing to accept relativized, perspectival conclusions rather than absolutely unqualified conclusions. It remains only to convince oneself that perspectivally realistic conclusions are not unacceptably relativistic, but, rather, the most reliable conclusions any human enterprise can produce.

Overlapping Theoretical Perspectives

At the end of chapter 3, I noted that the same object can often be observed from several different perspectives, such as a nearby galaxy observed by both optical and radio telescopes. This is indeed good evidence that there is "something" there, but that is scarcely knowledge in the objectivist sense. The knowledge we get comes from one perspective or another, not from no perspective at all. Multiplying perspectives does not eliminate perspectives.

Something similar applies to theoretical perspectives as well. The speed of light, for example, shows up in several different theoretical perspectives, including special relativity and quantum theory. Although the value of the constant, c, is determined by measurement, the *definition* is theoretical, the velocity of light in a perfect vacuum. It is a constant in an idealized model. Our best theories tell us that there are no perfect vacuums to be found anywhere in the universe. So-called empty space is full of all kinds of "space dust." If it were not, we could see a lot further with optical telescopes than we can in fact see.

Nevertheless, the fact that c shows up in several different theoretical perspectives is indeed significant. Moreover, models incorporating it do often exhibit an extraordinarily good fit with real physical systems. But being a component of several successful theoretical perspectives does not make something independent of all theoretical perspectives. So what we should conclude is that the speed of light is indeed fundamental to our most basic and experimentally best-tested theoretical perspectives on the physical world. From these perspectives, the velocity of light is indeed a fundamental constant of nature. That is as far as realism can go.

The Contingency Thesis Revisited

I think I have made a good case for thinking that scientific knowledge claims are perspectival rather than absolutely objective. It follows almost immediately that some contingency is always present in any science. That human observation is perspectival, a function of an interaction between the world and human cognitive capacities, seems to me indisputable. And that the instruments that now dominate scientific observation are similarly perspectival seems almost equally indisputable. They are designed to interact selectively with the world in ways determined by human purposes. That the development of theoretical principles and abstract models is in part a function of past knowledge and current interests and ideas is less obvious, but, on reflection, seems clearly correct. Still less obvious is the conclusion that all the previous contingencies are not eliminated by the application of sound experimental methods. Here we must first realize that experimentation can only involve the meshing of several perspectives, some observational and some theoretical. But a combination of perspectives remains perspectival, not objective in some stronger sense. And even within perspectives, the strongest possible conclusion is that some model provides a good but never perfect fit to aspects of the world. A supposed "absolute conception of the world" (Williams 1985) is unattainable.[47] This leaves us with a viable notion of scientific realism, but it is perspectival and contingent, not objective in some more absolute (metaphysical) sense.

One reason why the perspectival nature of existing human and instrumental observation seems undeniable is that we can understand the ways they are perspectival in terms of broader theoretical perspectives. We cannot do that for our current theoretical perspectives. Here there is need for what Kuhn (1962, chap. 1) called "a role for history." Looking back historically, we can examine and understand the perspectival nature of earlier theoretical perspectives.

The exemplary case is surely that involving the change first from an Aristotelian/Ptolemaic theoretical perspective to a Copernican/Newtonian perspective and then to a General Relativistic perspective. Here it is noteworthy that the Ptolemaic perspective could be maintained partly because observation of the heavens was limited to what could be seen with the naked human eye, although eventually instruments were designed that enabled people like Tycho Brahe to make quite accurate measurements of relative positions and angular distances among planets and stars. Although they came relatively late in the historical process of the Copernican Revolution, Galileo's telescopes made it virtually impossible to uphold the Ptolemaic perspective. The telescopic perspective meshed well with a Copernican perspective and much less well with the Ptolemaic. But from a Copernican perspective, we can well understand why earlier students of the heavens should have accepted the Ptolemaic perspective.

The Newtonian synthesis eventually enthroned a perspective including gravitational attraction at a distance that lasted some 200 years. By the middle of the nineteenth century, most physicists surely thought it nearly incomprehensible that the Newtonian perspective would ever be overthrown. But it has been dethroned. In a Newtonian framework, the motion of the Earth around the Sun is like that of a ball being whirled around on the end of a string. The force exerted through the string keeps the ball moving in a circle rather than flying off on a tangent. Disconcertingly, in the case of the earth, there is no string! In a relativistic framework, the motion of the earth is more like a ball rolling (frictionless) around the inside of a round cup. In this case, the force exerted by the surface of the cup keeps the ball from flying off. That the ball is attracted by something at the bottom of the cup is an illusion. We can understand the illusion of action at a distance from a relativistic perspective.[48] Maybe some physicists working on "theories of everything" have glimpses of yet another perspective.

A perspectival understanding of scientific claims ensures that there is some degree of contingency in all such claims. It thus opens up the *possibility* that contingency can be discovered through empirical (historical, sociological, psychological) research, but it does not ensure that the specific nature of the contingency can always be uncovered. Indeed, in the major historical examples of changes in theoretical perspective, the specific contingencies that sustained the earlier perspective became evident only from the vantage point of the later perspective. Often it is only from a new perspective that one can see, relative to that new perspective, where the earlier perspective was lacking.[49]

The Pessimistic Induction

Histories of the sort sketched above provide a quick argument against optimistic versions of objectivist realism. Nineteenth-century Newtonian physicists were surely as justified in thinking that they had discovered the objectively real structure of the world as any scientists could possibly be. Yet Newtonian gravitational theory has been abandoned as the accepted account of gravitational phenomena (though not for many applications for which it yields sufficiently accurate predictions). Can it be anything more than presentist hubris to think that we now have the objectively correct theory?

There is a stronger version of this argument that has been used against various versions of scientific realism. It goes like this. Most theories proposed in the past have proven to be false and were replaced by newer theories. By simple induction, it is overwhelmingly likely that our present theories will also prove to be false and be replaced. There is, therefore, no reason to take seriously claims about what is really in the world based on current scientific theories.[50] This argument assumes the older standard view that scientific theories are sets of straightforward empirical statements that must be either true or false. It fails to connect with a perspectival realism according to which the only empirical claims that may be counted as true or false are those about the fit of models to the world, and it is built into these claims that fit is not expected to be perfect. Relative to the scientific and experimental context, a claim of good fit may be rejected as false. But many such claims are justifiably taken to be true, so long as it is understood that "good fit" does not mean "perfect fit." And that understanding is built into perspectival realism.

Reflexivity Reconsidered

Any naturalistic approach to understanding science as a human activity is subject to the demand that the approach be applied reflexively to itself. Constructivist accounts of science have frequently had difficulties with the demand for reflexivity. Realist accounts have no such problems. The same holds, I think, finally, for perspectivally realist accounts of science. What I have offered may be regarded as a set of models of various scientific activities. And I have argued that these models exhibit a good fit to actual scientific practices. That, on my own account, is as much as anyone can do.

CHAPTER FIVE
PERSPECTIVAL KNOWLEDGE AND DISTRIBUTED COGNITION

Introduction

My argument that scientific knowledge is perspectival has proceeded largely by reflecting on the actual practice of science itself. To show that observational astronomy is perspectival, for example, I presented some aspects of contemporary astronomy and physics without appeals to other sciences or even to theoretical ideas from various areas of science studies. The argument that theorizing is perspectival did, however, appeal to a conception of modeling derived largely from the philosophy of science. Nevertheless, for the most part, my conclusion that scientific knowledge is perspectival stands on its own. Now, however, I will seek additional support for this conclusion from another source, *the cognitive study of science.*[1]

Many of the activities of scientists are directed toward the acquisition of new knowledge. Because acquiring knowledge is indisputably a cognitive activity, we can say that, to a large extent, science is a *cognitive* activity. It is thus an appropriate object for the application of the concepts and methods of the cognitive sciences. Here I will consider only one concept from the cognitive sciences, that of *distributed cognition.* Much of the cognitive activity of scientists, I will argue, involves the operation of distributed cognitive systems, most of which incorporate the sorts of instruments and models I have characterized as perspectival.

Distributed Cognition

The idea of distributed *processing* has long been a staple in computer science. Distributed *cognition* is an extension of the basic idea of distributed processing. How radical an extension is a contentious issue I will consider later in this chapter. For the moment I will begin with two of several contemporary sources of the notion of distributed cognition within the cognitive sciences.

The PDP Research Group

One source of the concept of distributed cognition comes out of disciplines usually regarded as being within the core of cognitive science: computer science, neuroscience, and psychology. This is the massive, two-volume *Parallel Distributed Processing: Explorations in the Microstructure of Cognition* produced by James McClelland, David Rumelhart, and the PDP Research Group, based mainly in San Diego during the early 1980s (McClelland, Rumelhart, and the PDP Research Group 1986). Among many other things, this group explored the capabilities of networks of simple processors thought to be functionally at least somewhat similar to neural structures in the human brain. It was discovered that what such networks do best is recognize and complete *patterns* in input provided by the environment. The generalization to humans is that much of human cognition is a matter of recognizing patterns through the activation of prototypes embodied in groups of neurons whose activities are influenced by prior sensory experience.

But if something like this is correct, how do humans do the kind of *linear* symbol processing apparently required for such fundamental cognitive activities as using language and doing mathematics?[2] Their suggestion was that humans do the kind of cognitive processing required for these linear activities by creating and manipulating *external representations*. These latter tasks *can* be done well by a complex pattern matcher. Consider the following, now canonical, example (1986, vol. 2, 44–48). Try to multiply two three-digit numbers, say 456 × 789, in your head. Few people can perform this simple arithmetical task. Figure 5.1 shows how many of us learned to do it.

This process involves an *external representation* consisting of written symbols. These symbols are manipulated, literally, by hand. The process

$$
\begin{array}{r}
4\,5\,6 \\
7\,8\,9 \\
\hline
4\,1\,0\,4 \\
3\,6\,4\,8 \\
3\,1\,9\,2 \\
\hline
3\,5\,9\,7\,8\,4
\end{array}
$$

Figure 5.1 The multiplication of two three-digit numbers using a traditional technique.

involves eye-hand motor coordination and is not simply going on in the head of the person doing the multiplying. The person's contribution is (1) constructing the external representation, (2) doing the correct manipulations in the right order, and (3) supplying the products for any two integers, which can be done easily from memory.

Notice that this example focuses on the *process* of multiplication, the task, not the product and not knowledge of the answer. Of course, if the task is done correctly, one does come to know the right answer, but the focus is on the *process* rather than the *product*. The emphasis is on the *cognitive system* instantiating the process rather than cognition simpliciter.

Now, what is the cognitive system that performs this task? Their answer was that it is not merely the mind-brain of the person doing the multiplication, nor even the whole *person* doing the multiplication, but the *system* consisting of the person *plus* the external physical representation. It is this whole system that performs the cognitive task, that is, the multiplication. The cognitive process is distributed between a person and an external representation.

Hutchins's Cognition in the Wild

Another source for the concept of distributed cognition within the cognitive sciences is Ed Hutchins's study of navigation in his 1995 book *Cognition in the Wild*.[3] This is an ethnographic study of traditional "pilotage," that is, navigation near land, as when coming into port.[4] Hutchins demonstrates that individual humans may be merely components in a complex cognitive system. No one human could physically do all the things that must be done to fulfill the cognitive task, in this case repeatedly determining the relative location of a traditional navy ship as it nears port. For example, there are sailors on each side of the ship who telescopically record angular locations of landmarks relative to the ship's gyrocompass. These readings are then passed on (e.g., by the ship's telephone) to the pilothouse, where they are combined by the navigator on a specially designed chart to plot the location of the ship. In this system, no one person could possibly perform all these tasks in the required time interval. And only the navigator, and perhaps his assistant, knows the outcome of the task until it is communicated to others in the pilothouse. Those recording the locations of landmarks have no reason to know the result of the process.

In Hutchins's detailed analysis, the social structure aboard the ship, and even the culture of the U.S. Navy, play a central role in the operation of this cognitive system. For example, it is important for the smooth operation of the system that the navigator holds a higher rank than those making the sightings. The navigator must be in a position to give orders to the others. The navigator, in turn, is responsible to the ship's pilot and captain for producing locations and bearings when they are needed. So the social system relating the human components is as much a part of the whole cognitive system as the physical arrangement of the required instruments. In general, it is the social arrangements that determine *how* the cognition gets distributed within the overall cognitive system.

One might wish to treat Hutchins's case merely as an example of "socially shared cognition" (Resnick, Levine, and Teasley 1991) or, more simply, *collective cognition*. The cognitive task—determining the location of the ship—is performed by a collective, an organized group, and, moreover, under the circumstances, could not physically be carried out by a single individual. In this sense, collective cognition is ubiquitous in modern societies. In many workplaces there are some tasks that are clearly cognitive and, under the circumstances, could not be carried out by a single individual acting alone. Completing the task requires coordinated action by several different people. Thus, Hutchins invites us to think differently about many common situations. Rather than simply assuming that all cognition is restricted to individuals, we are invited to think of some actual cognition as being distributed among several individuals.

Hutchins's conception of distributed cognition, however, goes beyond collective cognition. He includes not only persons but also instruments and other artifacts as parts of the cognitive system. Thus, among the components of the cognitive system determining the ship's position are the alidade used to observe the bearings of landmarks and the navigational chart on which bearings are drawn with a rulerlike device called a hoey. The ship's position is determined by the intersection of two lines drawn using bearings from the two sightings on opposite sides of the ship. So parts of the cognitive process take place not in anyone's head, but in an instrument or on a chart. The cognitive process is distributed among humans and material artifacts.

The standard view, of course, has been that things such as instruments and charts are "aids" to human cognition, which takes place only in someone's head. But the concept of an aid to cognition has remained vague. By expanding the concept of cognition to include these artifacts, Hutchins provides a clearer account of what things as different as instruments and charts have in common. They are parts of a distributed cognitive process.

Scientific Observation as Distributed Cognition

In chapter 3, I described the operations of the Hubble telescope system. I would now like to suggest that we think of the operation of the Hubble telescope as a complex version of the navigation system on Hutchins's traditional navy ship. Instead of maybe five people, we have fifty to five hundred people involved in any given extended observation. In place of simple eyepieces and charts, we have the highly sophisticated Hubble telescope and a chain of electronic transmissions ending at the Goddard Space Flight Center, where computers are programmed to generate elaborate images for easy viewing. Nevertheless, they are both, I suggest, distributed cognitive systems. The Hubble telescope system is designed for the cognitive task of producing images of objects in space, from nebulae in the Milky Way to galaxies in deepest space. The system includes not only sophisticated hardware, but also numerous humans organized into interconnected groups, each with specialized tasks to perform in order to produce the final output.

Although much smaller in scale, the same can be said about the various imaging facilities that are becoming commonplace in medical complexes. An MRI facility, for example, can be thought of as a relatively localized, but still quite complex, cognitive system for producing images of human brains and other organs. Once in place, such a facility can be operated by just a few technicians. The system becomes larger if one includes the scientists or medical specialists who then use the images for scientific or diagnostic purposes.

Why introduce the new notion of distributed cognition? Is this not merely a redescription of something we already understand in other terms? I will return to this question later. For the moment it is sufficient to my purposes that understanding modern scientific experimental facilities as distributed cognitive systems is compatible with understanding their outputs as being perspectival in nature. In this respect, a perspectival understanding of scientific knowledge and the cognitive study of science fit comfortably together and may therefore be mutually reinforcing.

Models as Parts of Distributed Cognitive Systems

Now I want to argue that it is not only observation and experimentation that can be understood as involving distributed cognitive systems, but theorizing as well. For me, of course, theorizing is largely a matter of constructing and reasoning about *models*. Here I will consider four types of models: diagrammatic, pictorial, physical, and abstract. The first three types are relatively easy to understand as being components in distributed cognitive systems. The fourth is more difficult to understand in this way, but can be so nonetheless.[5]

Reasoning with Diagrams

Rumelhart's example of multiplying two three-digit numbers by the standard pencil and paper method is already a kind of diagrammatic reasoning. Moreover, because a diagram is a kind of physical object, reasoning with diagrams is also a kind of reasoning with physical models. Nevertheless, diagrammatic reasoning is sufficiently interesting and important that it is worth considering in its own right.

Figure 5.2 provides an example of mathematical reasoning that makes essential use of diagrams. The two diagrams in figure 5.2 embody a famous proof of the Pythagorean Theorem. Note first that the areas of the two larger squares are identical and, indeed, equal to $(a + b)^2$. The area of the single smaller square in the left diagram is equal to c^2, while the areas of the two smaller squares in the right diagram are a^2 and b^2, respectively. All of the triangles must be of equal area, being right triangles with sides a and b, respectively. In the left diagram the area of the smaller square, c^2, is equal to the area of the larger square minus the areas of the four triangles. But similarly, in the right diagram the combined area of the two smaller squares, a^2 plus b^2, equals the area of the larger square minus the areas of the same four triangles. So it follows that $c^2 = a^2 + b^2$.

In this reasoning, crucial steps are supplied by observing the relative areas of squares and triangles in the diagrams. Very little propositional knowledge, for example, that the area of a right triangle is proportional to the product of the lengths of its two sides, is assumed. The logicians Jon Barwise and John Etchemendy (1996) call this reasoning "heterogeneous inference" because it involves *both* propositional and visual representations. The issue is how further to characterize this reasoning. Here there are a number of views currently held.

A very conservative view is expressed in the following passage written by a contemporary logician. Referring to a geometrical proof, he writes that the diagram "is only an heuristic to prompt certain trains of inference; . . . it is dispensable as a proof-theoretic device; indeed, . . . it has no proper place in the proof as such. For the proof is a syntactic object consisting only of sentences arranged in a finite and inspectable array" (Tennant 1986). This position simply begs the question as to whether there can be valid diagram-

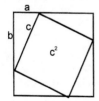
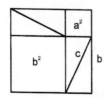

Figure 5.2 Diagrams used to prove the Pythagorean Theorem.

matic proofs or, indeed, even heterogeneous inference. Even granting that any diagrammatic proof is dispensable in the sense that the proof can be reconstructed as a syntactic object, it does not follow that diagrams can have "no proper place" in *any* legitimate proof. Only by assuming the notion of a proof to be essentially syntactic can one make this so.

Herbert Simon had a great interest in diagrams and took a more moderate position. He long ago (1978) claimed that two representations, for example, propositional and diagrammatic, might be "informationally equivalent" but not "computationally equivalent." By this he meant that there could be a method of computation that made the same information more accessible or more easily processed in one representational form rather than another. There remained, however, an important respect in which Simon's position was still relatively conservative, as illustrated in the following passage: "Understanding diagrammatic thinking will be of special importance to those who design human-computer interfaces, where the diagrams presented on computer screens must find their way to the Mind's Eye, there to be processed and reasoned about" (Simon 1995, xiii). Here Simon seems to be saying that all the cognition goes on in the head of the human, in view of "the mind's eye." Others are more explicit.

A common approach to diagrammatic reasoning in the artificial intelligence community is to treat diagrams simply as perceptual input that is then processed computationally (Chandrasekaran, Glasgow, and Narayanan 1995). Here, the most conservative approach is first to translate the diagram into propositional form and then proceed with processing in standard ways (Wang, Lee, and Hervat 1995). Figure 5.3 portrays diagrammatic reasoning as understood in standard artificial intelligence.

From the standpoint of distributed cognition, treating a diagram merely as input to a standard logic-based computer misses most of what is important about diagrammatic reasoning. What is interesting about diagrammatic reasoning is the *interaction* between the diagram and a human with a fundamentally pattern-matching brain. Rather than locating all the cognition in the human brain, one locates it in the system consisting of a

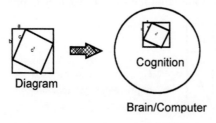

Diagram

Cognition

Brain/Computer

Figure 5.3 Diagrammatic reasoning in standard artificial intelligence.

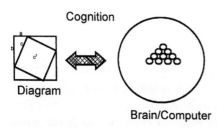

Figure 5.4 Diagrammatic reasoning as distributed cognition.

human together with a diagram. It is this system that performs the cognitive task, for example, proving the Pythagorean Theorem. The system can fairly easily perform this task, whereas the human alone might not be able to do it at all. A large part of what makes the combined system more powerful is that the geometrical relationships are embodied in the diagrams themselves. Extracting the desired relationships from the diagrams is far easier (especially for a pattern-matcher) than attempting to represent them internally. Figure 5.4 characterizes diagrammatic reasoning as a form of distributed cognition.[6]

From this perspective, there is at least one clear way of studying diagrammatic reasoning within the framework of cognitive science. This is to investigate the characteristics of diagrams that most effectively interact with human cognitive capabilities. Stephen Kosslyn's *Elements of Graph Design* (1994) provides a prototype for this kind of study. Unfortunately, he devotes only two pages explicitly to diagrams, but one of his examples shows the kind of thing one can do. The example concerns diagrams that show how pieces fit together in an artifact, as in figure 5.5. The relevant fact about human cognition is that humans process shape and spatial relationships using different neural systems, so the two types of information do not integrate well. The lesson is that, in a diagram, the representations of the pieces should be near to each other, as in the lower diagram.[7]

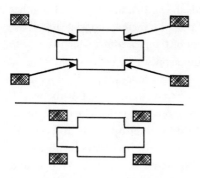

Figure 5.5 Bad (top) versus good (bottom) design of diagrams, based on cognitive principles.

Reasoning with Pictorial Representations

One cannot draw a sharp distinction between diagrams and pictorial representations. At one extreme we find highly idealized line drawings. At the other extreme are actual photographs, for example, of a volcano. In between we find a vast variety of things such as X-ray photographs of broken bones, telescopic pictures of distant nebulae, magnetic resonance images of brains, scanning electron microscope pictures of human tissues, magnetic profiles of the ocean floor, and so on.

Understanding pictorial reasoning as distributed cognition makes it possible to treat pictorial reasoning as more or less continuous with diagrammatic reasoning. The cognition is in the interaction between the viewer and the picture, as when geologists take magnetic profiles of the sea floor as indicators of sea-floor spreading (Giere 1996b). We need not imagine geologists as first forming a mental representation of the picture and then reasoning with it. They can reason directly with the external representation of the phenomenon.

Reasoning with Physical Models

From a distributed cognition perspective, there is again little fundamental difference between reasoning with diagrams and reasoning with physical models. A diagram is, after all, a kind of physical model, only two- rather than three-dimensional.

With some physical models one can work with the relevant parts of the world itself, the world being its own best model. A simple example is the problem of fitting slightly different lids on a number of slightly different screw-top jars.[8] Suppose a person cannot tell by simple inspection which lid fits which jar. There is, however, an obvious effective procedure for solving the problem without acquiring any new perceptual skills. Pick any lid and try it on the jars one at a time in any order until one finds the jar that fits. Then pick another lid and do likewise until all the jars are fitted with their appropriate lids. Here, one need only construct a representation of the procedure, which is easy to do, and then simply let the world itself guide one to the correct solution.

In a more normal case, particularly in the sciences, a physical model is constructed to be an *external representation* of some aspect of the world. An obvious example once again is Watson and Crick's original sheet metal and cardboard model of DNA held together by a pole and clamps. By fiddling with the physical model so as to fit the base pairs inside a double-helical backbone, they came up with the right structure. Watson and

Crick with their physical model in addition to Rosalind Franklin's X-ray pictures turned out to be a more effective cognitive system than Franklin herself with just her X-ray pictures and hand-drawn diagrams.

Physical models provide what is probably the best case for understanding reasoning with models as an example of distributed cognition. Here it is very clear that one need not be performing logical operations on an internal representation. It is sufficient to perform and observe appropriate physical operations on an external physical representation. The interaction between a person and the model is physical as well as perceptual.

Reasoning with Abstract Models

Abstract models provide what is probably the most difficult case for understanding reasoning with models as an example of distributed cognition. It is not clear in what sense an abstract model can be external. Nor is it clear how a person can interact with an abstract model. Yet many, if not most, models used in the sciences are abstract models. Think particularly of models in quantum physics or cosmology. So some account of reasoning with abstract models is needed.

The only alternative to regarding abstract models as *external* is to regard them as *internal*. There are, however, many reasons for rejecting this alternative. One reason is that we could not really refer to *the* model of anything because it is to be expected that every individual involved will have a somewhat different internal model. We would have to speak of A's model, B's model, and so on.[9] An even stronger reason is that it is highly implausible to suppose that all the details of the complex abstract models of contemporary science could be represented in the form of a mental model. Even experts, when asked to solve a difficult problem, typically proceed first to create an external representation of the problem in the form of diagrams or equations. The suggested account of this process is that the expert is using the external representations in order to *reconstruct* aspects of the abstract model relevant to the problem at hand. This no doubt requires some simple mental models. It also requires acquired *skills* in constructing the sorts of external models in question.

Yet even if we agree that abstract models are in some sense external, there remains a question of just what this means. This question threatens to lead us into the arid land of the philosophy of mathematics, where one worries about what numbers might be. I think we would do well to avoid this detour and take a safer route. As in our understanding of time, we traffic in abstract entities every day without worrying about what they are or how we interact with them. Consider plans and planning, well-known topics in cognitive science. Here abstract models are simply taken for granted. Suppose three friends are planning a party. The planned party is, at least in part, an abstract model of a party. It is assigned a date and time in the future and potential guests may even be designated in a written list. The party starts out as an abstract entity, a mere possibility, because it may in fact never materialize. I suppose almost everyone who has thought about the problem agrees that the ability to use language has a lot to do with the human ability to make plans, which is to create abstract entities, which we may take to be abstract models. This is enough for present purposes.

Here we already have the beginnings of one plausible answer to the question of how humans interact with abstract models. They do so by using language, both ordinary and highly technical. The three friends in my example build up their model of the party by talking about it. Moreover, they can reason about it as well, realizing, for example, that the number of potential guests has become too large for the intended space. It does not follow, however, that the possible party is itself in any way propositional, a mere linguistic entity. My three friends are not talking about what they are saying; they are talking about a possible party.

However important language might be in developing abstract models, it is not the only means for doing so. Diagrams, pictures, and physical models may also be used to characterize aspects of an abstract model. Watson and Crick's physical model of DNA, for example, also served the purpose of specifying some features of an abstract model of DNA, such as the pitch of the helix and the allowable base pairs. Other features of the physical model, such as being made partly of sheet metal, have no counterpart in the abstract model.

Even this rudimentary understanding of abstract models as abstract entities is enough to support our understanding of the development and use of abstract models as an instance of distributed cognition. Traditional scientists sitting alone with pencil and paper are already distributed cognitive systems. They interact physically with their diagrams and equations and, thereby, abstractly with an assumed more complex abstract model. Two such scientists working at a blackboard exploring the features of an abstract model are more obviously a distributed cognitive system. Now imagine the more likely contemporary scene in which a group of theoreticians are working on networked computers developing a set of equations that define a complex model. Here one has a conception of a distributed cognitive system consisting of humans and computers producing an emerging abstract model.

Computation in Scientific Distributed Cognitive Systems

Having said that much cognitive activity in science is distributed, one can still go on to ask for a further characterization of these cognitive processes. And here the standard cognitive science account might still be applied, namely, that the overall process is *computational*. Indeed, one could say that, since its inception in the 1970s, modern cognitive science has operated within a paradigm which understands cognition as computation.[10] And, strictly speaking, computation means there is a languagelike symbolic representation that is manipulated according to explicit rules.

Consider again the Compton Gamma Ray Observatory (CGRO), which I would take to be an exemplar of a distributed cognitive system in science. Now there clearly were many parts of this system that were computational in the strict sense. There were standard digital computers all along the path from onboard data acquisition to the final generation of an electronic image at a facility such as Goddard. Indeed, it is astounding the extent to which, since the 1950s, the computer has changed the way science is done. One could say that computers have changed the whole culture of science in fundamental ways.[11] Nevertheless, it does not follow that the overall cognitive process of acquiring knowledge through the practice of modern science is to be characterized as computational.[12]

Consider first the initial stage detector, which operated on the principles of Compton scattering of gamma rays by electrons. This is a physical process. It is doing something, but I see little justification for calling what it is doing *computational*. There are no symbolic representations and no rules for manipulating representations involved in the interaction between gamma rays and electrons. Those came later in the data-acquisition and data-processing stages. I am inclined to follow the principle that, where there are no symbolic representations and no rules, there is no computation. A physical process, yes. Computation, no.

Consider one of the simplest possible physical processes, a ball held steady and then released from a height, h, above the ground. Assuming nothing interferes, it will, of course, fall to the ground. The elementary Galilean equation for the height, y, of the ball (or its center of gravity) as a function of the time, t, since its release is, as every student of elementary physics knows, $y = h - 1/2gt^2$, where g is the gravitational constant for the Earth. Now, if the ball were actually calculating its height as a function of time, it could not be doing so according to this equation, which ignores air friction and the fact that g is not constant with respect to height. It would have to be calculating the "real" function, which no one knows how to write down in complete detail. But to say it is calculating the "true" height as a function of time seems to me pretty nearly vacuous. If it were calculating, what else could it calculate? No. Far better, I think, not to ascribe any computations at all to the falling ball. It just falls. It is *we* who have devised various equations to compute (always and only) approximate values for the parameters of its fall.[13]

Agency in Scientific Distributed Cognitive Systems

Besides the computers and physical processes, the CGRO also involved many people. It has been central to the standard paradigm in cognitive science that humans, insofar as they are cognitive, are computational. But should human cognition, after all, be understood as being computational? This is one of the deepest theoretical issues in cognitive science, and far too big an issue to take up here. Instead, I will focus on a related issue, the location of *agency* in distributed cognitive systems.[14]

Extending Agency

Many of those who have been developing notions of distributed cognition, or related notions, have been tempted to suppose that this way of thinking requires extending our conceptions of knowledge, mind, and even conscious agency, thus attributing to systems like CGRO as a whole attributes heretofore restricted to human agents. Here I will argue against such extensions, concluding that it is preferable to restrict applications of such cognitive attributes as knowing and conscious intentional agency to the human components of distributed cognitive systems. I begin with some prominent examples of those who would extend agency to distributed cognitive systems as a whole.

Foremost in this category is Hutchins himself, who argues that we should regard *mind* as extended beyond the human body. There is mind at work, he has claimed, on the visible surface of the chart as the navigator and his assistant point to representations of various landmarks and decide which landmarks to use for the next sightings. Apparently Hutchins thinks that these decisions are to some extent literally made on the chart rather than just in the heads of the navigator and his assistant.[15]

Andy Clark's (1997) spirited advocacy of cognition as being distributed focuses on what I would call *locally distributed cognition*. A person working with a computer would be a paradigm example. The person with a computer can perform cognitive tasks, such as complex numerical calculations, that a person alone could not possibly accomplish as accurately or as fast, if at all.

When it comes to arguing for extending the concept of *mind* beyond the confines of a human body, however, he invokes a more primitive example, that of a man with a defective memory who always keeps on hand a notebook in which he records all sorts of things that most of us would just remember (1997, 213–18). Clark claims that the person's mind should be thought of as including the notebook. For this person, the process of remembering something important typically involves consulting his notebook. The notebook is part of his memory. A major component of Clark's argument for this position is that for someone else deliberately to damage the notebook would be equivalent to assaulting the person. The notebook is as crucial to this man's normal cognitive functioning as is the left part of his brain.

In her latest book (1999), Karin Knorr-Cetina, one of the founders of social constructivism in the sociology of science, argues that *different* scientific fields exhibit *different* epistemic cultures.[16] Her first and most extensive case is high-energy physics (HEP), in particular, experiments done between 1987 and 1996 at the European Center for Nuclear Research (CERN). The scale of this laboratory is suggested by the fact that CERN's Large Electron Positron Collider, located on the border between France and Switzerland, was 27 kilometers around. This collider is now being replaced by a Large Hadron Collider (LHC) coupled with a very large detector called ATLAS. The ATLAS detector itself is 44 meters wide, 22 meters high, and weighs seven thousand tons. The ATLAS project involves hundreds of scientists, technicians, and other support personal.

My view, of course, is that, like CGRO or the Hubble telescope, the laboratory at CERN should be thought of as a large distributed cognitive system. Knorr-Cetina, in fact, indirectly suggests this approach. In at least a half-dozen passages, she uses the term "distributed cognition" to describe what is going on in a HEP experiment. Here are two examples: "the subjectivity of participants is . . . quite successfully replaced by something like distributed cognition" (1999, 25); "Discourse channels individual knowledge into the experiment, providing it with a sort of distributed cognition or a stream of (collective) *self-knowledge*, which flows from the astonishingly intricate webs of communication pathways" (1999, 173). Her uses of the expression "distributed cognition" are almost always qualified with expressions such as "something like" or "a sort of," and there is never any further characterization of what distributed cognition might be.

Perhaps Knorr-Cetina's most provocative idea is "the erasure of the individual as an epistemic subject" in HEP. She claims that one cannot identify any individual person, or even a small group of individuals, producing the resulting knowledge. The only available epistemic agent, she suggests, is the extended experiment itself. Indeed, she attributes to the experiment itself a kind of "self-knowledge" generated by the continual testing of components and procedures and by the continual informal sharing of information by participants. In the end, she invokes the Durkheimian notion of "collective consciousness." Speaking of stories scientists tell among themselves, she writes: "The stories articulated in formal and informal reports provide the experiments with a sort of consciousness: an uninterrupted hum of self-knowledge in which all efforts are anchored and from which new lines of work will follow" (1999, 178). And on the following page, she continues: "Collective consciousness distinguishes itself from individual consciousness in that it is public: the discourse which runs through an experiment provides for the extended 'publicity' of technical objects and activities and, as a consequence, for everyone having the possibility to know and assess for themselves what needs to be done" (1999, 179).

Here Knorr-Cetina seems to be assuming that, if knowledge is being produced, there must be an epistemic subject, the thing that knows what comes to be known. Moreover, knowing requires a subject with a mind, where minds are typically conscious. Being unable to find a traditional epistemic subject within the organization of experiments in HEP, she therefore feels herself forced to find another epistemic subject, settling eventually on the experiment itself as the epistemic subject.

Problems with Extending Agency

The first thing to realize is that applying concepts associated with human agency (mind, consciousness, intentionality) to extended entities involving both humans and artifacts, or to inanimate entities themselves, is a matter of fairly high level *interpretation*. One cannot even imagine an *empirical* test of Clark's claim that his man's notebook is part of the man's mind. Even if a court of law determined that stealing the man's notebook is as serious a crime as assaulting him bodily, that would not show that the notebook should be included as part of his mind. Similarly for the navigator's chart on Hutchins's ship or for an experiment at CERN. If we are to adopt these interpretations, it can only be because they provide theoretical benefits of some sort, benefits that cannot be obtained without these innovations. My view is that these extensions do not provide theoretical advantages for the study of science. On the contrary, they introduce a host of theoretical problems that confuse more than enlighten. We are theoretically better off rejecting these supposed innovations.[17]

The culture in scientifically advanced societies includes a concept of a human agent. According to this concept, agents are said to have minds as well as bodies. Agents are conscious of things in their environment and self-conscious of themselves as actors in their environment. Agents have beliefs about themselves and their environments. Agents have memories of things past. Agents are capable of making plans and sometimes intentionally carrying them out. Agents are also responsible for their actions according to the standards of the culture and local communities. And they may justifiably claim to know some things and not other things. There is much more that could be said about the details of this concept of an agent, but this is enough for my purposes here.

The traditional difficulty with the ordinary conception of human agency is that it seems to presuppose a notion of freedom of choice and action that is incompatible with a scientific (naturalistic, materialistic) understanding of humans as biological organisms. To stay within a naturalistic framework, I am willing to grant that the underlying causes of our actions are largely hidden from us and that the subjective impression that one can in a strong sense freely cause things to happen (particularly one's own bodily motions) is an illusion.[18] So I am willing to regard the ordinary concept of an agent as an idealized model, like that of a point mass in classical mechanics. Such things do not physically exist, but it is a useful model nonetheless. The same could be true of our ordinary notion of human agency. Like idealized models in the sciences, it has proven useful in organizing our individual and collective lives. Indeed, our systems of morality and justice are built upon it.

There is no doubt that *some* extensions of concepts originating with humans beyond the bounds of biological agents are natural, even helpful. Memory is a prime example. Modern civilization, as well as modern science, would be impossible without various forms of recordkeeping that are usefully characterized as external memory devices. The modern computer hard drive, for example, is a recent, powerful memory device. The organization of such devices, the software, so to speak, is equally important. Thus, a person with a computer, like Clark's man with his notebook, is usefully thought of as a cognitive system with an effective memory much more powerful than a system consisting of a person alone, without any such devices. But even here it is pushing the applicability of the concept of memory to say that the extended system as a whole "remembers" something, as opposed simply to having the capacity to store large amounts of information.

The situation is similar for "collective cognition," which I would characterize as cognition distributed among persons only, without bringing in artifacts. The Hubble system, like the system at CERN, or even the system aboard Hutchins's ship, includes numbers of individuals organized so as to accomplish tasks necessary for the whole system to function properly. Collectively they make the system work. To understand how the members of the groups collectively make the system work, it is not necessary, and, I think, definitely unhelpful, to introduce the concept of a super agent. Our ordinary concept of a human agent, plus concepts from such fields as psychology, sociology, organization theory, and anthropology, is sufficient to provide as good an account as we need of how all the individuals interact to produce the final cognitive output. The basic notion is simply that the individuals, acting together, make the system work.[19]

Problems arise, however, when one attempts to extend concepts like consciousness, intention, belief, knowledge, and responsibility to extended entities. These concepts are all bound up in the more general concept of "mind." Consider again the Hubble telescope. As a distributed cognitive system it extends at least from the telescope in orbit through a series of intermediaries to the Space Telescope Science Institute in Maryland. If one adds the Abell 1689 cluster of galaxies used as a gravitational lens, the system extends 2.2 billion light-years out into space. Are we to say that its mind extends from the telescope in orbit to Maryland, or 2.2 billion light years out into space? Do minds operate at the speed of light? Just how fast do intentions propagate? Is the Hubble system as a whole epistemically (as opposed to just causally) responsible for the final conclusions? Did the system as a whole expect to find galaxies as far as 13 billion light years away? Did it then believe it had found them? These questions don't make much sense. We should not have even to consider them. So we should resist the temptation to ascribe full agency, including having a mind, to distributed cognitive systems.

Distributed Cognition and Human Agency

One might now wish to raise similar questions about the initial conception of distributed cognition. Does not this notion invite similar unanswerable questions? I think not. The word *cognition* was part of the English language before the field of cognitive science was invented, but I think always somewhat of a specialists' term.[20] We are thus freer to develop it as a technical term of cognitive science. The fundamental notion, I think, is not so much that of distributed cognition as that of a *distributed cognitive system*. A distributed cognitive system is a system that produces cognitive outputs, just as an agricultural system yields agricultural products. The operation of a cognitive system is a cognitive process. There is no difficulty in thinking that the whole system, no matter how large, is involved in the process. But there is no need also to endow the system as a whole with other attributes of human cognitive agents. So thinking of cognition as distributed throughout the system need not raise untoward questions.

But what makes the output of the system a *cognitive* output? Here I think the only basis we have for judging an output to be cognitive is that it is the kind of output we recognize as the result of human cognition, such as a belief, knowledge, or a representation of something. This means that the surest way to guarantee that the output of a distributed cognitive system be cognitive is that there be a human agent in the output stage of the system. This is currently the case for existing scientific distributed cognitive systems. Whether there could be a distributed cognitive system with no human components is then left as an open question.

Actually, as has long been taken for granted in the general science studies community, a claim does not count as *scientific* knowledge until it is publicized and accepted by the relevant scientific community. In John Ziman's (1968) terms, scientific knowledge is "public knowledge." This means there will be a short period of time when members of the research group who reached consensus on the result can each claim personally to know the result even though the result does not yet count as scientific knowledge. This implies that the cognitive system that produces scientific knowledge should really be taken to be a whole scientific community, including things such as the institutions that make publishing possible.[21] So scientific distributed cognitive systems turn out, finally, to be quite heterogeneous systems with fuzzy boundaries.

If we still need a way to characterize scientific knowledge in general, my suggestion would be to *depersonalize* the characterization, so that we would say things like "This experiment leads to the scientific conclusion that . . ." and "It has been shown scientifically that . . ." These more impersonal forms of expression free us from the need to find a special sort of epistemic subject for accepted scientific knowledge.

Distributed Cognitive Systems as Hybrid Systems

Bruno Latour (1993) has popularized the notion of hybrid systems consisting of combinations of humans and nonhumans, all called "actants." I would also like to say that distributed cognitive systems are hybrid systems, but retain the genuine differences between humans and nonhumans. In more detail, I think distributed cognitive systems include at least three distinct kinds of systems: physical, computational, and human. It is the humans, and only the humans, that provide intentional, cognitive agency to scientific distributed cognitive systems. We need not extend our notions of cognitive agency to include other components of these distributed cognitive systems. Restricted to humans, our ordinary notions of human agency will do. The ultimate justification for this theoretical choice is that it provides a productive framework that all contributors to science studies, including historians, philosophers, psychologists, sociologists and anthropologists, can share.[22]

Why Distributed Cognition?

I will now return to the question of whether introducing the notion of distributed cognition is necessary. Can we not say everything we want to say using more standard notions? Is this not just a redescription of situations we already know how to describe? Yes and no. When I claim that CERN is a large distributed cognitive system, it is not as if I have myself gone to Switzerland and discovered things going on that Knorr-Cetina somehow missed. Not at all. What I have done is reconceptualize the activities she describes in other terms. I claim that this new way of thinking about the activities provides a richer understanding of what is going on.[23] In particular, we can now say that aspects of the situation that seemed only social are also cognitive. That is, aspects of the social organization determine how the cognition gets distributed. So here we have a way of overcoming some of the opposition that many have felt between social and cognitive understandings of science.[24]

Introducing the notion of distributed cognition into studies of science also has the advantage of connecting such studies with broader movements within the cognitive sciences. Although I have focused on just two sources of recent interest in distributed cognition, the idea is much more widespread, as one can learn from Andy Clark's *Being There: Putting Brain, Body, and World Together Again* (1997). Lucy Suchman's *Plans and Situated Actions* (1987) and *The Embodied Mind* by Varela, Thompson, and Rosch (1993) are among earlier influential works. A common theme here is that the human brain evolved primarily to coordinate movements of the *body*, thereby increasing effectiveness in activities such as hunting, mating, and rearing the young. Evolution favored cognition for effective action, not for contemplation. *Cognition is embodied.* This point of view argues against there being a single central processor controlling all activities and for there being many more specialized processors. Central processing is just too cumbersome. Thus, an emphasis on distributed cognition goes along with a rejection of strongly computational approaches to cognition. In fact, some recent advocates of dynamic systems theory (Thelen and Smith 1994) have gone so far as to argue against there being any need at all for computation as the manipulation of internal representations.

Clark, in particular, invokes the notion of "scaffolding" to describe the role of such things as diagrams and arithmetical schemas. They provide support for human capabilities. So the above examples involving multiplication and the Pythagorean Theorem are instances of scaffolded cognition. Such external structures make it possible for a person with a pattern-matching and pattern-completing brain to perform cognitive tasks it could not otherwise accomplish.

The most ambitious claim made in behalf of distributed cognition is that *language* itself is an elaborate external scaffold supporting not only communication, but thinking as well (Clark 1997, chap. 10). During childhood, the scaffolding is maintained by adults who provide instruction and oral examples. This is analogous to parents supporting an infant as it learns to walk. Inner speech develops as the child learns to repeat instructions and examples to itself. Later, thinking and talking to oneself (often silently) make it seem as though language is fundamentally the external expression of inner thought, whereas, in origin, just the reverse is true. The capacity for *inner* thought expressed in language results from an internalization of the originally *external* forms of representation. There is no separate "language of thought." Rather, thinking in language is a manifestation of a pattern-matching brain trained on external linguistic structures (see also Bechtel 1996).

This distributed view of language implies that cognition is not only embodied, but also *embedded* in a society and in a historically developed culture. The scaffolding that supports language is a cultural product (Clark 1997).[25] This view has recently been developed under the title of "cognitive linguistics" or "functional linguistics" (Velichkovsky and Rumbaugh 1996). Some advocates of cognitive linguistics emphasize comparative studies of apes and humans. Tomasello (1996, 1999, 2003) claims that a major difference between apes and human children is *socialization*. Of course, there are some differences in genotype and anatomy, but these are surprisingly small. The bonobo named Kanzi, raised somewhat like a human child by Duane Rumbaugh and Sue Savage-Rumbaugh (Savage-Rumbaugh 1993), is said to have reached the linguistic level of a two-year-old human child. Tomasello argues that what makes the difference between a natural and an acculturated chimpanzee is developing a sense of oneself and others as *intentional* agents. Natural chimpanzees do not achieve this, but acculturated ones can to some extent. The result, of course, is that natural chimpanzees are incapable of developing the sort of cumulative culture that language makes possible (Tomasello 1999).

From the standpoint of computational linguistics, there has always been a question of how the neural machinery to support the necessary computations could possibly have evolved. From the standpoint of cognitive linguistics, this problem simply disappears. Language is not fundamentally computational at all, but the product of a pattern-matching neural structure, which biologically could evolve, supported by an elaborate scaffolding of social interaction within an established culture.

Distributed Cognition and Perspectival Knowledge

I think that most scientific knowledge now being produced is the product of distributed cognitive systems. Such systems incorporate various sorts of humanly produced artifacts, both material and abstract. And these artifacts typically incorporate a built-in perspective on the world. Beginning with an understanding of scientific knowledge as produced by distributed cognitive systems, therefore, one comes quickly to the conclusion that scientific knowledge is perspectival.

Nor is this a new development. With hindsight, we can see that newly created distributed cognitive systems played a large role in the Scientific Revolution. Cartesian coordinates and the calculus, for example, provided a wealth of new external representations that could be manipulated to good advantage. And new instruments such as the telescope and microscope made possible the creation of extended cognitive systems for acquiring new empirical knowledge of the material world. There remains, of course, the historical question of how all these new forms of distributed cognitive systems happened to come together when they did, but their impact is undeniable. And the history of science from then on is filled with an ever-expanding range of new distributed cognitive systems providing new perspectives on the world.

NOTES

CHAPTER ONE SCIENTIFIC KNOWLEDGE

1. I deal only with Western culture (primarily North American and Western European) for the simple reason that, because it is my culture, it is what I know best. And this is where most of my colleagues and members of my audience reside. It is better to state this up front than to pretend to universality. Others are invited to adapt what I say here to their own circumstances.

2. The classic work in this genre is Rachel Carson's *Silent Spring* (1962).

3. Here, the classic work is Betty Friedan's *The Feminine Mystique* (1963). See also Cowan (1983).

4. References to some of the works that provoked the backlash against social and cultural studies of science can be found in the critiques from the scientific side (Gross and Levitt 1994; Gross, Levitt, and Lewis 1996; Koertge 1998). For a reply to some of these attacks, see Ross (1996). A balanced presentation of the science wars from a British perspective has been compiled by Labinger and Collins (2001).

5. Late-twentieth-century debates about scientific realism are covered in several book-length defenses of various forms of scientific realism (Leplin 1997; Niiniluoto 1999; Psillos 1999). I have myself recently (2005) contributed to this literature. The development of social constructivist sociology of science and its interaction with philosophy and the philosophy of science during the second half of the twentieth century are nicely traced by the intellectual historian John Zammito (2004). For another attempt at reconciliation from a feminist point of view, see Longino (2002).

6. Comparisons between my scientific perspectivism and the perspectivisms of earlier thinkers, such as Leibniz, Kant, and Nietzsche, would be a worthwhile enterprise, but I am not prepared to pursue that task here. I respect the history of philosophy too much to engage in historical comparisons without a thorough knowledge of the philosophy and also of the culture of the time in question. For Kant's perspectivism, see Palmquist (1993), and for Nietzsche, see Hales and Welshon (2000).

7. What I am here calling "common-sense realism" corresponds to what the cognitive linguist George Lakoff (1987, 158–60) calls "basic realism" and Hilary Putnam (1999), following William James, calls "natural realism." Michael Devitt (1991), whose naturalistic views I find generally appealing, seems to me to move too easily from what he also calls "common-sense realism" to something close to an objectivist scientific realism. I will have more to say about common-sense realism in chapter 2.

8. See also Weinberg's 1992 book, *Dreams of a Final Theory: The Scientist's Search for the Ultimate Laws of Nature.* The subtitle nicely highlights the object-

ivist idea that there are such things as "the ultimate laws of nature" to be discovered.

9. Such sentiments are common in the more popular scientific literature, as illustrated by the following remark by Brian Greene (1999, 183): "A physicist's dream is that the search for the ultimate answers will lead to a single, unique, absolutely inevitable conclusion." For at least hints of an objective realist position in molecular biology, there is Walter Gilbert's (1992) famous remark: "Three billion bases of sequence can be put on a single compact disk (CD), and one will be able to pull a CD out of one's pocket and say, 'Here is a human being; it's me!'"

10. See the books referred to in note 5, also Leplin (1984).

11. For Putnam's original statement of his "proof" of the incoherence of metaphysical realism, see his 1981 book (ch. 2 and appendix). A good summary of the ensuing debate over Putnam's argument appears in Lakoff (1987, chap. 15). For more recent assessments, see Hale and Wright (1997) and Forrai (2001).

12. See, in particular, chapter 11, "What about God?" in *Dreams of a Final Theory* (1992). Moreover, Weinberg should not need reminding that, at the end of the nineteenth century, physicists were as justified as they could possibly be in thinking that classical mechanics was objectively true. That confidence was shattered by the eventual success of relativity theory and quantum mechanics a generation later.

13. Perspectival realism is a further development of what I earlier (1988, 1999a) called constructive realism. My initial thoughts about the possibility of a perspectival realism benefited from discussions with my former student Laura Rediehs and from her dissertation (Rediehs 1998).

14. My rejection of objectivist realism is similar to Nancy Cartwright's recent rejection of what she calls "fundamentalism" (1999), the view that everything that happens is governed by a universal law of nature. For my assessment of the similarities and differences in our views, see "Models, Metaphysics, and Methodology" (Giere, forthcoming).

15. My evidence for this claim is a Library of Congress catalog search as suggested by Hacking (1999, 24).

16. The only basis I can imagine for a clear denial of this claim is that our social order was dictated by some god. Even sociobiologists and evolutionary psychologists admit that our biological nature permits a range of social arrangements from which some are selected by contingent historical and social circumstances.

17. As I review these words, the President of the United States has just requested that the United States Congress initiate the process of amending the U.S. Constitution to declare that, in the United States, marriage applies only to the union of one man and one woman. That this action is even possible provides unambiguous evidence of the socially constructed nature of marriage, and thus of the categories of husband and wife.

18. The title of her first book (1981) was *The Manufacture of Knowledge: An Essay on the Constructivist and Contextual Nature of Science*. Her most recent book, *Epistemic Cultures: How the Sciences Make Knowledge* (1999) is relatively free of constructivist proclamations.

19. Collins has recently (2004) completed an extensive new work on contemporary gravity research. He has also considerably softened his constructivist views. See his contributions to Labinger and Collins's *The One Culture* (2001).

20. See Galison (1997, 540–45) for a detailed analysis of the experiments that Pickering claims show how physicists "tuned" their instruments in conformity to the new theory, thus choosing the new physics over the old rather than experimentally verifying the new. Galison finds no mention of the theory in any documents, published or unpublished, by the experimental groups. Moreover, in interviews he conducted, the physicists in question insisted they were not thinking about the theory at the time or, in some cases, did not even know much about it. This sort of historical research, and not proclamations of the objectivity of science, seems to me exactly the right kind of response to constructivist studies, meeting them on their own grounds.

21. There is general agreement that the two theories agree on all implications that are empirically detectable. Thus, no experiment could provide decisive evidence in favor of one over the other. But they are conceptually very different. So it is difficult to argue that they are mere notational variants. On the standard interpretation, quantum phenomena are inherently probabilistic. The Bohm theory is deterministic. Standard quantum mechanics is incomplete in the sense that it provides no dynamic account of measurement processes. The Bohm theory eliminates this "measurement problem." It is fully dynamic. But it is dramatically nonlocal, which is to say, it invokes action at a distance. So does the standard theory, though in more subtle ways.

22. If these assertions sound like a religious creed, that is deliberate. The quoted lines are preceded by the sentence: "Our faith in these principles is partly religious, partly pragmatic."

23. In discussions with Werner Callebaut (1993, 184), Knorr-Cetina used the terms "teleological realism" and "epistemological realism" to make just the point of this paragraph.

24. This point was made perhaps most forcefully by Laudan (1981).

25. Yet it seems that Collins and Yearley (1992) did just this, advocating a "social realism" as the basis for science studies.

26. This, I take it, is the lesson of Steve Woolgar's (1988) reflexive turn in the sociology of scientific knowledge.

27. Indeed, I would subscribe to the original principles of the Strong Programme in the sociology of science, namely, that explanations of episodes in science should be causal, impartial, symmetric, and reflexive (Bloor 1976). These principles basically prescribe a naturalistic approach to studying science, with which I agree. I disagree only with the excessively sociological *implementation* of these naturalistic principles by practitioners of SSK. My own implementation is more "cognitive" and less "social," but makes room for the "social" as well. I am also concerned with the instrumental effectiveness of experimental methods in helping scientists choose better fitting models over less well fitting models, something sociologists of science contest.

28. Here I am adopting the terminology introduced into the philosophy of science by van Fraassen in his recent book *The Empirical Stance* (2002).

29. But see the recent book by Golding (2003).

30. I have developed the connection between naturalism and Pragmatism in somewhat more detail in the essay "Naturalism and Realism" (Giere 1999a).

31. Feminists may appreciate the parallel between my perspectivism and Haraway's (1991).

32. For an overview of the philosophy of science in the period 1940–60, see my "From *Wissenschaftliche Philosophie* to Philosophy of Science" (1996a). The standard source for structural-functional sociology of science is Merton (1973).

33. Philosophers, of course, debated the limits of objectivist scientific claims. Should they extend beyond objects observable by means of unaided human perception? Similarly, many physicists and some philosophers of science questioned the extent to which an objectivist account could be applied to the quantum realm. In neither of these cases, however, were the worries generated by concerns regarding the actual cognitive capacities of scientists or the actual social environments in which they worked.

34. The reference here, of course, is to the various social constructivists discussed above.

35. For those familiar with the history of philosophy, my project is roughly the reverse of Kant's. Kant took Newtonian physics to be in some sense both universally and necessarily true of the world. He set himself the task of showing how such knowledge is possible. His philosophical Copernican Revolution was to conclude that such knowledge is possible because the truths of classical physics are imposed by the constitution of any rational mind. My main task is to show how contingency in scientific knowledge is possible. My conclusion is based on the actual practices of the sciences and the empirically determined cognitive capacities of actual human agents.

36. This view has been stressed by Alan Sokal (1998).

CHAPTER TWO **COLOR VISION**

1. Contemporary philosophical interest in color vision began with Hardin's (1988) *Color for Philosophers*. For a good selection of subsequent papers by analytic philosophers see Byrne and Hilbert (1997a).

2. There are also receptors called rods (again, for their shape), which are sensitive only to relative light and darkness. Rods are overall more light sensitive than cones, so that, as the level of light decreases, one's vision becomes more and more simply black and white. For my purposes here, we can take the activity of rods for granted. As will be noted later, interaction between rods and cones, particularly in lower mammals, can provide some ability to discriminate objects on the basis of spectral differences in reflected light.

3. In the nineteenth century, Hermann von Helmholtz championed a trichromatic theory of color vision according to which all colors resulted from combinations of excitation of three color-sensitive receptors, red, green, and blue. Ewald Hering advocated an opponent-process theory with a red-green system and a separate blue-yellow system. The modern dual-process version of the opponent-process

theory, dating only from 1957 (Hurvich 1981), provides a synthesis of these two theories in which differences in three component intensities produce a four-component result. For a brief history, see Palmer (1999, 107–12).

4. Here we have a straightforward scientific explanation of a phenomenon much discussed in the history of both philosophy and psychology: there seem to be no such colors as a greenish red or a reddish green. Or, as it is often stated in the philosophical literature, nothing can be red and green all over. This is not a synthetic a priori truth, nor a truth of grammar, as Wittgenstein might have said (Westphal 1987), but a straightforward scientific conclusion.

5. In addition, pigments in the retina actually adapt to light level, becoming more sensitive in low light and less so when the light is more intense (Hurvich 1981, chap. 15).

6. Note that the chromatic-response function is asymmetric. If we suppose that the philosophers' mythical person with an "inverted spectrum" has a reversed chromatic-response function, this fact could be discovered by the very techniques used to determine chromatic-response functions in the first place. If it is simply stipulated that such a person have a normal chromatic-response function, then the example becomes physically much less plausible and, therefore, less interesting.

7. Byrne and Hilbert (2003, n. 3) provide five examples of similar quotations from other standard works in color science.

8. The comparison between colors and the rising of the sun is also made by Boghossian and Velleman (1989).

9. In terminology fashionable in some philosophical circles, one would say that the color of an object *supervenes* on the physical properties of the incident light and the surface of the object.

10. Hilbert (1987) argues that the colors of objects are objectively (physically) real just because the surface spectral reflectance of an object together with the known chromatic-response function for normal humans determines the chromatic experience a normal human will report. He calls this "anthropocentric realism" for colors. The real question, however, is whether, in the absence of human visual systems, there would be, for example, any physical basis for identifying as red the hodgepodge of surfaces with spectral reflectances that existing humans experience as being red. And even Hilbert seems to admit that the answer to this question is no, when he says, for example, "In a world in which there were no human beings, . . . [t]here would be color, but since color plays little role in the causal laws that govern a world without perceivers, it would tell us little of interest in such a world" (1987, 14–15). I would amend the final clause just slightly by saying that colors, as such, play *no* role in the causal structure of a world without perceivers. Surface spectral reflectances, yes; colors, no. To think otherwise is to think that an arbitrary disjunction of physical properties is a legitimate physical property even though such a property has no significance whatsoever in any physical theory.

11. My objections to identifying colors with surface spectral reflectances apply also to versions of the so-called dispositional account of colors that goes back at least

to Descartes, who, in his *Principles of Philosophy* wrote: "We therefore must on all counts conclude that the objective external realities that we designate by the words *light, colour, odour, flavour, sound,* or by the names of tactile qualities such as *heat* and *cold,* and even the so-called *substantial forms,* are not recognizably anything other than the powers that objects have to set our nerves in motion in various ways, according to their own varied disposition" (Anscombe and Geach 1954, 234). This view is echoed by Newton, Boyle, and Locke (Thompson 1995, chap. 1). The hypothesized dispositions ("powers"), however, have no independently identifiable physical basis. They are just the dispositions of a disparate set of objects that happen to interact optically with the contingently evolved sensory capacities of humans in humanly identifiable ways.

12. In this section, I rely heavily on Thompson (1995, chap. 4).

13. Cases of total achromatopsia are delightfully described by Oliver Sacks (1997).

14. Here I draw heavily on Jacobs (1993).

15. Jacobs (1993, 423) reports that of 4,321 species of mammals recorded as existing in 1981, only about 100 had been studied for evidence of color vision.

16. In linguistic terms, one tends not to think of relational predicates when thinking of properties, but they are commonplace, as in "x is taller than y" or "x is married to y." The relational predicate for color would be something like "x appears yellow to y" where y might be "a normal trichromat" or "a dichromat," etc.

17. Byrne and Hilbert (2003, n. 3) cite passages in several other texts on vision or color science that clearly express a subjective view of colors. In each case, however, a closer reading reveals a more relational (I would say "perspectival") point of view. Sekuler and Blake (1985, 181) are cited as saying, "[O]bjects themselves have no color. . . . Instead, color is a psychological phenomenon, an entirely subjective experience." Yet the very next sentence continues in a more perspectival vein: "Objects *appear* colored because they reflect light from particular regions of the visible spectrum. But even that is not enough. In order for an object to appear colored, light reflected from that object must be picked [up] by the right sort of eye or nervous system." Kaiser and Boynton (1996, 486) are reported as asserting that "color is ultimately subjective." Yet on the very same page they state that color is a *relation* between spectral reflectances and color sensations: "[T]he relation between the spectral distribution of light reflected to the eye from an artificially isolated specimen and the color sensation that it elicits can be reasonably well understood." Finally, Fairchild (1998, xv) is quoted as saying that "without the human observer there is no color." But this is true also on a relational account of color vision. Fairchild goes on to list a dozen well-know observations, such as: "There are no colors described as reddish-green or yellowish-blue." He concludes: "None of the above observations can be explained by physical measurements of materials and/or illumination alone. Rather, such physical measurements must be combined with measurements of the prevailing viewing conditions and models of human visual perception in order to make reasonable predictions of these effects" (xvi). In short, color perception is relational.

18. In a recent article in the *American Journal of Physics* entitled "Human Color Vision and the Unsaturated Blue Color of the Daytime Sky" (Smith 2005), a computer scientist explains why the daytime sky is blue by invoking the *interaction* between sunlight scattered by the Earth's atmosphere and the human visual system. The spectrum of visible light reaching the earth's atmosphere is relatively flat, peaking around 500 nm (green) and falling off slowly in both directions. But due to Rayleigh scattering of light by air molecules (mostly nitrogen and oxygen), the intensity of visible sunlight reaching the surface of the Earth peaks at the violet end of the visible spectrum and falls off as the fourth power of the wavelength for longer wavelengths. So why does the sky not appear more violet than blue? Because, given the chromatic-response function of the human visual system, the "spectral irradiance" (the counterpart of spectral reflectance for transmitted light) of the transmitted sunlight is metamerically equivalent to a pure blue light (475 nm) mixed with pure white light, which is experienced as light blue. So even the claim that the sky is blue is not an absolutely objective truth. Rather, the sky appears blue to normal human trichromats.

19. Paul Churchland has recently (2005) published a delightful and philosophically interesting paper on the neuroscience of color vision.

20. I am also presuming that the causal mechanisms of the visual systems are effectively deterministic. That is, even though the interactions between light and the visual system may be indeterministic at the level of light quanta and visual pigments, the numbers of such individual interactions involved in ordinary vision are so large that the behavior at the macro level is highly constrained. More generally, of course, there is no incompatibility if two identical, but fundamentally stochastic, systems sometimes produce different outputs for the same inputs.

21. I say not *visually* detected because one can, of course, feel heat generated by electromagnetic radiation that is not visible.

22. I presume this term is borrowed from Husserl, or at least related to Husserl's use of this term, but I am not in a position to explore this possible connection.

23. One of the most criticized of Thomas Kuhn's suggestions was that "after a revolution scientists are responding to a different world" (1962, 110).

24. For a well-documented presentation of this point of view, see Wilson (1992).

25. A representative example of current philosophical work in the philosophy of color is provided in Byrne and Hilbert's introduction to their authoritative anthology of recent papers on the subject (1997a). They begin by saying: "There is a central concern shared by nearly all the papers collected here (and most contemporary philosophical discussions of color), namely, to answer the following questions: Are physical objects colored? And, if so, what is the nature of the color properties? These questions form the problem of *color realism*" (xi; original italics). They proceed to outline four answers to these questions. "*Eliminativists* answer no to the first question. They think that although physical objects seem to be colored, this is no more than a peculiarly stable illusion. . . . *Dispositionalists* say that the property green (for example) is a disposition to produce certain perceptual states: roughly, the disposition

to *look green*. . . . *Physicalists* claim that colors are physical properties (for instance that green is a certain property of selectively reflecting incident light). . . . *Primitivists* agree with the physicalists that objects have colors. . . . But they also hold that colors are sui generis, and so they deny, in particular, that colors are identical to physical properties." Regarding dispositionalists, in particular, their "basic claim . . . is something of the following form": "The property blue = the disposition to cause in perceivers *P* in circumstances *C*, visual experiences of kind *K*." A major question, then, is whether there is any plausible candidate for kind *K* other than the obviously circular "experiences of seeing blue."

The questions that form "the problem of color realism" are peculiarly philosophical, not to say just plain peculiar. Common sense and common language answer unequivocally, "Yes (what a silly question)!" Yet the science of color shows, I think, that the scientifically correct answer is not simply either yes or no. Rather, humans see objects as being colored because of a complex interaction between the way various objects reflect incident light and the way the human visual system processes that reflected light. Not only are the questions peculiar, so are the answers. The dispositionalist answer above is in the form of an analytical definition. That is a form peculiar to analytic philosophy. One will not find such definitions in textbooks of vision science. Once one knows the science, what can be the point of producing such definitions? Given the peculiarity of the question, it is hardly surprising that there is so little agreement among philosophers as to which such definition is correct.

The philosophy of color is also much concerned with the *experience* of seeing colors. Here is something on that subject, again from Byrne and Hilbert's introduction: "Because a visual experience may be veridical or illusory, we may speak of visual experiences *representing* the world to be in a certain way. For example, when you look at a tomato, your visual experience represents the world as containing, inter alia, a round red object in front of you. That is, the *representational content* (content, for short) of your experience includes the proposition that there is something red and round before you. . . . Let us say that a visual experience is *red-representing* just in case it represents that something is red" (xii–xv). Visual experiences do more than represent, they also have distinctive subjective quality. About this quality, it is said: "Experiences of red objects resemble one another in a salient phenomenological respect. . . . Let us say that a *red-feeling experience* is an experience of the kind picked out by the following examples: the typical visual experiences of ripe tomatoes, rubies, blood, and so forth." The question is now raised whether red-feeling visual experiences are *necessarily* also red-representing experiences. The answer to this question is said to depend on the truth of the following proposition: "For all possible subjects *S*, and possible worlds *w*, *S* is having a red-feeling experience in *w*, if and only if *S* is having a red-representing experience in *w*." The truth of this proposition turns on the *possibility* of there being a subject having a red-feeling experience that is not also a red-representing experience, or vice versa. Such a possibility is said to be provided by a version of Locke's well-known hypothetical example of a person who experiences the spectrum in an inverted form, so that this person would have a chromatic experience like that of seeing a Marigold when in fact looking at a Violet. Since such a person would be having a red-representing experience while having a yellow-feeling experience, red-representing experiences are *not necessarily* connected with red-feeling experiences. And this result is said to have important implications for many issues in the philosophy of color.

This issue tells us more about current fashions in the philosophy of mind and language than it does about color vision. It involves peculiarly philosophical notions of experience and representation. And it concerns not empirical findings about things in the world, but logically (or conceptually) necessary connections among concepts. One cannot imagine a color scientist being concerned with generalizations over all actual subjects, let alone all *possible* subjects and in all *possible worlds*. That is not how genuine science is done.

During the first quarter of the twentieth century, before philosophy and psychology became fully separate disciplines, the philosophy curriculum contained courses on such topics as memory and perception. Such courses continued to be taught in some philosophy departments even after the split. By the last quarter of the century, such courses had pretty much disappeared. Memory and perception were by then subjects for psychology, or cognitive science more generally. I think that is now the case for color vision. Given the current advanced state of the science of color vision, there is simply no need for a separate technical specialty, the philosophy of color vision.

26. As I understand it, Stroud argues that one could not intelligibly have the experience of seeing that a ripe lemon is yellow unless ripe lemons are objectively yellow apart from the existence of any perceivers. I think this claim injects far too much unwarranted metaphysics into the commonsense view that most of us can see that ripe lemons are yellow.

27. Although his bibliography contains reference to several books by philosophers (not color scientists) from which he could have learned some of the relevant color science, Stroud's book shows little appreciation for modern color science. By contrast, references to early modern thinkers, such as Locke and Hume, abound.

CHAPTER THREE SCIENTIFIC OBSERVING

1. See Daston (1992), and Daston and Galison (1992).

2. According to NASA's "History of the Hubble Space Telescope" (www.NASA.gov), such a telescope was proposed already in 1923 by the German scientist Hermann Oberth. Regrettably, it now (spring 2005) looks like NASA will let the Hubble die a natural death rather than attempt an earlier scheduled maintenance mission by the space shuttle.

3. The transition from theoretical speculation to instrumental application is a theme of Hacking's *Representing and Intervening* (1983) and also of my *Explaining Science* (1988).

4. The following is based on information obtainable through the NASA Web site or, more directly, at wwwgro.unh.edu/comptel.

5. Note how similar these considerations are to those emphasized by Galison (1997) in his studies of high-energy physics.

6. This information comes from NASA press release 97-83, dated April 28, 1997.

7. Comparing human perception with instruments also suggests that much philosophical thought about human visual perception is confused. Direct realists argue that we perceive objects themselves. Representational realists argue that we experience not the object itself, but a mental representation of the object. For instruments, the direct realists are closer to the mark. Instruments clearly do not form representations of objects, which they then detect. Instruments interact directly with objects in the world. They typically *produce* representations of the objects in question. But the representation is a *product* of the interaction between instrument and object, not part of the process by which the instrument detects aspects of the object. It must always be remembered, however, that an instrument interacts only with *aspects* of an object, for example, its emitted gamma-ray spectrum. Observation is thus always mediated; not, however, by a representation, but by the perceptual apparatus of the observer. Or, as I would sum it up, an observer, whether a human or an instrument, can interact with an object only from the observer's own particular perspective.

8. My understanding of imaging techniques in neuroscience has benefited from many conversations with my former student, Pauline Sargent, and from her Ph.D. dissertation (Sargent 1997).

9. Although not a major focus of my project here, it is worth emphasizing that CAT and the other imaging technologies discussed below only became possible with the development of high-speed computers. It is only CAT, the first widely used of these technologies, that explicitly acknowledges this dependence in its very name.

10. Medically, of course, a distinctive feature of computer assisted tomography is that it is *invasive* in a potentially destructive way. It requires sending ionizing X-rays through the tissue, and this can have undesirable consequences, such as the formation of cancers.

11. There are many sources explaining the operation of MRI machines. Here, I follow Toga and Mazziotta (1996, chap. 9).

12. Because MRI operates at radio frequencies, which are relatively harmless to both humans and animals, it provides an ideally noninvasive method of imaging all parts of the body. This partly explains its popularity and the recent great investment in improving MRI methods.

13. The advantage of detecting differences in oxygenated versus deoxygenated blood is that the blood is acting as an endogenous contrast agent, with all the advantages that typically go along with noninvasive methods.

14. I cannot resist quoting Galileo's description in *The Starry Messenger* on this point: "I have observed the nature and the material of the Milky Way. With the aid of the telescope this has been scrutinized so directly and with such ocular certainty that all the disputes which have vexed philosophers through so many ages have been resolved, and we are at last freed from wordy debates about it. The galaxy is, in fact,

nothing but a congeries of innumerable stars grouped together in clusters. Upon whatever part of it the telescope is directed, a vast crowd of stars is immediately presented to view. Many of them are rather large and quite bright, while the number of smaller ones is quite beyond calculation" (Drake 1957, 49). We may forgive Galileo his objectivist enthusiasm.

CHAPTER FOUR **SCIENTIFIC THEORIZING**

1. Philosophers of science, in particular, may remember the heated debates of the 1970s over the so-called theory ladenness of observation. I am not here invoking that notion, nor its opposite, an "observation ladenness of theory." These debates seem to me to have been based on several deep misunderstandings of scientific practice. One is that there is a general distinction between the language of theory and the language of observation that cuts across all sciences. Such a distinction might make sense if one thinks that scientific observation is homogeneous, as it would be, for example, if what we observe were only sense data, or even just what can be detected with unaided human perceptual capabilities. But in scientific practice there is no such uniform observational base. What counts as observation in science depends on what can be detected using whatever technology is available, and the technology keeps changing. In particular experiments, one can distinguish a theory being tested from the data used to test the theory, but this distinction is always local. There was a time when images on photographic film served as data for the existence of theoretically conjectured X-rays ("X" for unknown!). Now, detected X-rays are used as data regarding theoretically conjectured objects in the distant universe. For a good account of the discovery and impact of X-rays, see Kevles (1997). I have elaborated the point of this paragraph in "Scientific Realism: Old and New Problems" (2005).

2. A second misunderstanding was that, if the meaning of observational terms were determined by the meaning of theoretical terms, observation could no longer provide an independent basis for testing theoretical claims. This worry presumed a kind of meaning holism that today is untenable. In designing the COMPTEL gamma ray telescope, for example, astrophysicists had theoretical reasons for wanting to detect gamma rays of around 1.8 MeV coming from the center of galaxies, such as our own Milky Way. These theoretical concerns in no way prevented them from designing COMPTEL so that it would be unlikely to indicate the existence of a high flux of 1.8 MeV gamma rays if there were in fact few to be detected.

2. The conception of theorizing to be presented is derived from the model-based account of theories I have advocated in many earlier publications (Giere 1988, 1999a).

3. As suggested by Hacking (1983) more than two decades ago. See also my "How Models Are Used to Represent Reality" (2004a).

4. I have elsewhere examined the use of pictures, graphs, and diagrams as representational media (Giere 1996b).

5. This picture applies only to mature theories, such as quantum mechanics or evolutionary biology, in which there are recognized principles. As has been emphasized by Cartwright (1999) and Morgan and Morrison (1999), much science is done using

principles drawn from several different disciplines or using no principles at all. In such cases there is a less tight connection between principles and models.

6. I take abstract entities to be human constructions, the ability to create such constructions being made possible by symbolic artifacts such as language and mathematics. But abstract models are definitely not to be identified with linguistic entities such as words or equations. Any particular abstract model can be characterized in many different ways. Nor should abstract models be thought of as merely formal. They are created already interpreted.

To make the thought of abstract objects less mysterious, it may help to consider historical novels, for example, Tolstoy's *War and Peace*. In writing this novel, Tolstoy created an elaborate possible world. I could call it an elaborate model of a possible world. It is not, however, what I will be calling a "representational" model. It is no criticism of the novel that the characters it describes never existed in the real world. On the other hand, it does, and was intended to, represent *kinds* of human situations, attitudes, and emotions.

7. In the old debate over whether Newton's laws should be regarded as empirical claims or definitions, I turn out to be much closer to those who argued for the definitional point of view.

8. But see chapter 5 for a little more about language in general.

9. I distinguished interpretation and identification in *Explaining Science* (1988, 74–76). These two activities were conflated in the old Logical Empiricist notion of correspondence rules.

10. Thus, the arrow in figure 4.1 down to models from principles and specific conditions definitely does not indicate deduction. Again, as emphasized by Cartwright and others, constructing models is a complex activity that frequently includes a variety of approximations and simplifications.

11. Representational models are also to be distinguished from the "instantial" or "interpretive" models of logic and formal model theory (Giere 1999b).

12. This point has been effectively argued by Mauricio Suarez (2003). See also Teller (2001).

13. An appropriate slogan might be: No representation without representers.

14. The version of the semantic view of theories developed by Suppes (1957, 1967, 1969), Sneed (1971), and many others since (Baltzer and Moulines 1996) is based on the notion of a set-theoretical predicate.

15. It is for this reason that I would not follow Niiniluoto (1987, 1999) or Kuipers (2000) in trying to develop a general notion of truthlikeness or truth approximation for individual statements.

16. Here I assume that the partial model in question does not have a unique extension to a complete model. The possibility of there being such an extension seems most implausible.

17. For applications of these ideas in chemistry, see the work by Christie (1994) and Ramsey (1997).

18. See Teller (2001) for an extended discussion of the shortcomings of what he calls "the perfect model model" of science.

19. This view has been developed most prominently by Nancy Cartwright (1983, 1989, 1999) and also recently by Paul Teller (2004a, 2004b).

20. A drawback of using the term *fit* to describe the relationship between a representational model and a real system is that well-entrenched expressions such as "goodness of fit" suggest what is meant is just fit between measurable quantities and parameters in the model. When I speak of the "fit" of a model to the world, I do not intend any restriction to features of the model corresponding only to measured features of the real system. In short, I do not mean merely that the model "fits the data" but that it fits the whole system under investigation.

21. As I understand them, it is part of the job of a naturalistic philosophy of science, and science studies more generally, to construct what would ordinarily be called a theory of science. This pretty much requires some regimentation in the usage of existing terms as well as the introduction of some new concepts. It is obviously desirable to follow widely accepted usage as much as possible. In this spirit, I appropriated the term *principle* for things that within the sciences themselves are often referred to as "principles." But compromises are necessary, and so, as argued above, I prefer not appropriate the terms *theory* and *law* for my own account of scientific practice. The different things often called "theories" and "laws" mostly find a place in my account, just not quite the place they occupy in other accounts.

22. I have already said more in the title essay of *Science without Laws* (Giere 1999a). See that essay also for additional references relevant to my remarks in this section.

23. Here one may recall the story by Borges (1954) in which the cartographers of a fictional land set out to construct a map of their land on a scale of one to one. As they complete their project, the people of the land begin to move onto the new territory.

24. Even topological maps, of course, embody many truths, such as that in the D.C. subway system there is at least one major station that serves four different lines.

25. The subway system in Madrid, Spain, has one line that is just a circuit. The map for this line consists of an elongated rectangle with rounded corners and station names spaced at equal intervals around the perimeter. The shape of the map is determined primarily by the fact that it must both be easy to read and fit into the space between the top of the doors and the roof of the car. There is little need for riders to know either the geographic shape of the circuit or the relative distances between stops.

26. Australians are fond of displaying "upside down" maps with Australia at the top.

27. For an amusing introduction to cartography, see Monmonier (1991).

28. I have discussed this example at length in *Explaining Science* (1988, chap. 8).

29. In the seventeenth century, artisans produced matched pairs of globes, one being a globe of the earth, the other a representation of the celestial sphere. Several pairs of such globes can be seen at the Boerhaave museum in Leiden, The Netherlands.

30. Dennis Overbye, "New Map Unveiled of Universe at Start of Time," *New York Times*, February 11, 2003. For more information, see the NASA Web site at map.gsfc.nasa.gov/.

31. Here I ignore the complication that the Earth is not exactly a sphere, but an oblate ellipsoid of revolution, somewhat flattened at the poles.

32. In 1998 the National Geographic Society endorsed the Winkel Tripel Projection as providing a somewhat better compromise between preserving shape and areas for land masses than that of the Robinson map.

33. Margaret Masterman is famous for arguing at length that Kuhn used the term, *paradigm*, "in not less than twenty-one different senses" (1970, 61). Dudley Shapere's (1966) well-known review of *Structure* focused harshly on Kuhn's ambiguous use of this term.

34. Regarding this controversy, see, for example, Andersen (2001), Bird (2000), and Hoyningen-Huene and Sankey (2001).

35. In his later writings, Kuhn (2000) talked more about scientists learning a new language than about translating from one language into another.

36. This is also the conclusion of a relatively neutral observer, John Zammito (2004), an intellectual historian who has critically chronicled developments in the history, philosophy, and sociology of science throughout the second half of the twentieth century.

37. In the 1970s, Hilary Putnam (1975) and Saul Kripke (1980) developed a causal theory of reference to circumvent what they took to be the problem of incommensurability of meaning. The idea, derived from Quine, is that what matters about scientific language is not meaning, but reference. They argued, therefore, that there can be continuity of reference through changes in meaning. Reference is fixed by causal interaction with the real stuff. So the reference of the term *water* was fixed by our (presumably) English-speaking ancestors who were literally in physical contact with water running in rivers. Modern science, they said, has shown us that water is really H_2O, so our ancestors were in fact referring to H_2O even though they lacked our conception of chemical elements. Moreover, even though it took scientific research to learn the chemical composition of water, the resulting claim "Water is H_2O" is, according to Kripke's possible worlds semantics, a necessary truth, true in all possible worlds. I agree with the few brave critics, such as Paul Churchland (1989, chap. 13), that all this is an unnecessary solution to a manufactured problem. It embodies both an objectivist understanding of scientific claims and a naive view of actual science. What Robin Hood experienced as water in the Sherwood Forest was surely a solution containing all kinds of mineral salts and organic compounds, such as those from rotting vegetation, or worse. People living in London surely experienced an even more complex solution. And those who experienced ocean water were in contact with a solution containing lots more salts. Even today, almost nobody has ever been in causal contact with pure H_2O. Much of our water now includes chlorine, to say nothing of PCBs. Thus, taken seriously, the causal theory of reference fails even to yield the historical continuity it was designed to provide. It should be noted that

Putnam (1981) later explicitly rejected what he then dubbed "metaphysical realism." For recent careful analyses and ultimate rejections of the Putnam-Kripke approach, see Bird (1998, chap. 3) and LaPorte (2004).

38. Hacking (1991) credits John Venn as the originator or the term "natural kind."

39. For a classic constructivist account of categorization into kinds, see Bloor (1982).

40. The best overall reference on the "species problem" in the twentieth century (and earlier too) is Hull's *Science as a Process* (1988). Dupré (1993) provides a more radical critique of essentialism in biology.

41. I should note that I do not mean to deny both scientific and everyday common sense by claiming, as discussed in chapter 1, that the planets did not exist before the Scientific Revolution. Rather, the already known bodies (Earth, Sun, Moon, Mercury, Venus, Mars, Jupiter, Saturn) were simply *reclassified*. Roughly speaking, in moving from a Ptolemaic to a Copernican model of the "universe," the Earth and the Sun exchanged places. Whereas, in the Ptolemaic model, the Sun moved in an orbit between Venus and Mars around a central Earth, in the Copernican model, the Earth moves in that orbit around a central Sun. So the Earth became classified, along with Mercury, Venus, Mars, Jupiter, and Saturn, as a planet of the Sun, with the Moon remaining a satellite of the Earth and the Sun being the central body. Thus, before the Scientific Revolution, Mars was not *classified* as a planet of the Sun (nor was anything else), but it was no part of the Copernican Revolution that Mars had not previously been a planet. On the contrary. It is only the exaggerated rhetoric of the early constructivist movement in the sociology of science that made this seem like an issue requiring serious attention.

42. Because theoretical kinds are defined by theoretical principles, and principles are human constructs, my account of theoretical kinds ends up being on the constructivist side of the constructivist/objectivist divide—if one divides up the possible approaches to kinds that way, which I obviously do not. Here I have been most concerned to distinguish my view from objectivist views.

43. This includes my own *Explaining Science* (1988).

44. For a good introduction to this literature, see van Fraassen (1980) and the critical essays focusing on van Fraassen's views in Churchland and Hooker (1985). Given that I have adopted major features of van Fraassen's approach to scientific theories, it is no accident that my perspectival realism is in some ways similar to his constructive empiricism. The major difference is that van Fraassen limits the empirical to what ordinary humans can observe with their unaided senses. I impose no such demarcation between the empirical and the theoretical. To do so seems to me contrary to scientific practice. Rather, for me, which features of a model we take to have counterparts in the real world depends on what differences in the world can be reliably detected by whatever means, however remote. I am, however, inclined to draw a line at differences in the model whose postulated real-world counterparts cannot be detected because, in this case, we would have no empirical grounds for any claims about these differences. Here I part company with those, such as Churchland

(1989, chap. 8), who would appeal to superempirical virtues, such as simplicity or elegance. It does not seem possible to establish even just a reliable correlation between any superempirical virtue and good fit between models and the world. Thus, if a theoretical difference cannot be empirically detected, one should, I think, remain agnostic about the existence of such a difference in the real world.

45. I have provided more extensive accounts of such decision making in "Constructive Realism" (Giere 1985) and *Explaining Science* (Giere 1988, chap. 6)

46. Note that the decision to regard the results as positive or negative evidence for good overall fit of the models to the subject matter is based on judgments regarding the reliability of the overall experimental procedure and not on judgments about the probability of claims about good overall fit. This account thus bypasses standard antirealist arguments based on the fact that, because probability varies inversely with content relative to any observational evidence, claims about fit regarding only observable parameters are always more probable than claims regarding overall fit. In fact, I do not believe there is any empirical basis for making the kind of probability judgments this argument assumes, but this is a much disputed issue in the philosophy of science.

Also, note again that my account of experimental procedures does not presume any global distinction between what is observable and what is not. There is only a local distinction between models of the data and representational models determined by the particular experimental context. Thus, the data may be described in terms of the observed relative density of 1.8 MeV gamma rays coming from various regions of the Milky Way. This description would, of course, be based on the observed relative density of recorded events of a specified character in various regions of the detector, but there is no scientific rationale for insisting on taking a model of these events as the model of data to be compared with the representational models of nuclear physics and the dynamics of star formation. In any case, the density of recorded events, which is what is pictured in plate 6, is itself already a model of the data produced from recordings of individual events. There is much processing, both physical and computational, between individual event recordings and images like that of plate 6. The important scientific task is to develop both the models of the data and the representational models in such a manner that the data models can be meaningfully compared with some aspects of the representational models. In terms of the many intermediate levels implicit in figure 4.1, there is no one fixed level at which models of the data must meet representational models, just so long as they do meet somewhere between the principles (if such there be) and models of data. Or, again, in terms of perspectives, the task is somehow to create an overlap in relevant observational and theoretical perspectives. It does not matter at what level the overlap occurs, just so long as the observational perspective provides a contingent basis for evaluating the general fit of the representational models.

47. I say "unattainable" rather than "does not exist," because the former leaves open whether a supposed "absolute conception of the world" is possible or not, thereby "delivering us from metaphysics" (van Fraassen 1980, 69).

48. Of course, in a relativistic perspective, there is no cup either, just the curvature of space-time itself.

49. This seems to be the basic insight behind Paul Feyerabend's (1962) insistence on the value of proliferating theories and Miriam Solomon's (2001) much more recent advocacy of what she calls "Whig Realism."

50. In the philosophy of science, this sort of argument goes under the name "the pessimistic induction." For a recent extensive discussion, see Psillos (1999, chap. 5).

CHAPTER FIVE PERSPECTIVAL KNOWLEDGE AND DISTRIBUTED COGNITION

1. The basic structure of this chapter follows that of my "Scientific Cognition as Distributed Cognition" (2002c).

2. Note that this is just the reverse of the situation with standard, von Neumann processors, which are very good at performing rule-governed manipulations on symbolic structures but very poor at pattern recognition.

3. I wish to thank Linnda Caporael and Nancy Nersessian for first introducing me to the work of Ed Hutchins and others developing the concept of distributed cognition in the cognitive science community. Nersessian, in particular, shared with me her own grant proposal on cognitive and social understandings of science and even dragged me off to hear Ed Hutchins at a meeting of the Cognitive Science Society.

4. Note that the human cognitive skills employed in this task have been rendered obsolete by the invention of global positioning systems.

5. This section draws heavily on my "Models as Parts of Distributed Cognitive Systems" (Giere 2002b).

6. Figure 5.4 is oversimplified in that it pictures a brain in isolation from a body and, thus, without any means of physically interacting with the diagram.

7. Although not exactly to the point of this chapter, I would like to mention that diagrams, apparently because they are processed in similar ways by most humans, provide a means for effective communication across disciplines. This phenomenon, as I noted in *Explaining Science* (Giere 1988, chap. 8), was much in evidence during the 1960s revolution in geology.

8. I think I owe this example to Andy Clark (1997), but I have been unable retrieve an exact page number.

9. I do not want to overemphasize the uniqueness of abstract models or sets of models. As Hull (1988) has argued in more traditional terms, it is difficult to identify a unique set of models that could be designated as *the* models of evolutionary theory. Different groups of scientists, and sometimes different individuals within groups, deploy different sets of models under the general rubric of evolutionary models. From this fact it does not follow, however, that the models are mental rather than abstract entities.

10. Even Hutchins regards himself as working within this paradigm. He describes the distributed cognitive system of navigation aboard his navy ship as a computa-

tional system. It computes the location of the ship. He does this knowing full well that many of the instruments that are part of this system, such as the navigator's chart, are analogue devices. He thinks we should extend the concept of computation to include such systems. As will become clear, my inclination is not to extend the concept of computation and thus to admit that there are important cognitive processes, or parts of cognitive processes, that are not computational.

11. With few exceptions (e.g., Suppe 1989; Humphreys 2004), philosophers of science have been slow to recognize how fundamentally the use of computers has changed the nature of science. Historians of science, such as Galison (1997), and sociologists of science, such as Collins (1990), have been much more sensitive to this development.

12. This section draws heavily on my "The Role of Computation in Scientific Cognition" (Giere 2003).

13. These considerations, however, are not completely decisive. Influential people in physics and computer science (notoriously, Wolfram 2002) argue that the whole universe is one gigantic computer. They claim that space and time are discrete, so that it makes sense to say that the universe computes its next state (or the probability thereof) from its previous state. Now I agree with those (Lakoff and Johnson 1980, Lakoff 1987, Johnson 1987) who argue that language is deeply metaphorical. Nevertheless, in the sciences, the appropriateness of particular metaphors may be contested. In the sixteenth and seventeenth centuries, natural philosophers were much impressed with the elaborate clockworks then being produced. The universe, they concluded, was a gigantic clockwork. Newton's physics was taken as providing legitimization for the literal truth of this claim. We now know this was mistaken. The digital computer is, without doubt, the dominant technological innovation of the current epoch. It is not surprising that it is taken as more than a mere metaphor, not only for the mind (Turkle 1984), but for the universe as a whole. However compelling it now seems, I think this metaphor will turn out to be no better than the clockwork metaphor. I suspect that its attractiveness is at least partly rooted in a mistaken desire for a single, overarching explanation for everything. It is also a prime example of what Gerd Gigerenzer (2000) calls the move "from tools to theories," as, indeed, is the cognition as computation paradigm itself.

14. This section draws on my "The Problem of Agency in Scientific Distributed Cognitive Systems" (Giere 2004b), which also includes consideration of the views of Bruno Latour.

15. I have not been able to find these claims about mind in *Cognition in the Wild*. I did, however, personally hear Hutchins make these claims in a richly illustrated plenary lecture at the Annual Conference of the Cognitive Science Society in Boston, MA, July 31–August 2, 2003.

16. The following discussion of Knorr-Cetina's views draws on my "Distributed Cognition in Epistemic Cultures" (Giere 2002a).

17. For a detailed critique of the idea of extended minds from the standpoint of analytical philosophy of mind, see Rupert (2004).

18. The latest and best word I know on this topic is due not to a philosopher but to a psychologist, Daniel Wegener (2002).

19. Here I am not claiming that the scientists who interpret the final images need to know every detail of the system or could successfully perform most of the tasks required to keep the system operating. Knorr-Cetina is surely correct about the distribution of expertise needed to operate large-scale experimental systems. I question only her attribution of agency and self-consciousness to the operations of the system as a whole.

20. Among Anglo-American philosophers, for example, *cognitive* has typically been associated with *rational* or *reasons*. Thus, a philosopher would distinguish between a person's cognitive grounds (reasons) for a particular belief and mere causes of that belief.

21. In terms of Anglo-American philosophical epistemology, the justification of scientific claims, both personal and collective, is based on prior judgments regarding the *reliability* of the system producing the claims. This would include the system of peer review almost universal in the sciences.

22. I imagine that most historians of science and technology, for example, would have little difficulty with notions of distributed or collective cognition. But most, I think, would balk at the notion of extended minds and be very suspicious of talk about collective consciousness.

23. Dare I say that introducing the concept of distributed cognition creates a new *perspective* from which to view the operation of scientific facilities such as CERN and the Hubble telescope?

24. This point is developed in more detail by Giere and Moffatt (2003). This article focuses on two studies by Bruno Latour.

25. Interestingly, an externalized and socialized view of language was advocated in the 1920s by the Soviet psychologist Lev Vygotsky (Vygotsky 1962, 1978; Wertsch 1985). A product of the intellectual ferment inspired by the Russian Revolution, Vygotsky explicitly appealed to Marxist thought for the idea that the key to understanding how language developed lay not so much in the mind as in society. Nevertheless, his major 1934 book (translated in 1962 as *Thought and Language*) was suppressed within the Soviet Union from 1936 to 1956 and has only since the 1960s received notice in the English-speaking world.

REFERENCES

Adams, Robert M. 1987. Flavors, Colors, and God. In *The Virtue of Faith and Other Essays in Philosophical Theology*. New York: Oxford Univ. Press.

Alpers, Svetlana. 1983. *The Art of Describing*. Chicago: Univ. of Chicago Press.

Andersen, Hanne. 2001. *On Kuhn*. Belmont, CA: Wadsworth.

Anscombe, E., and P. T. Geach, eds. 1954. *Descartes: Philosophical Writings*. Edinburgh: Nelson.

Baltzer, W., and C. U. Moulines, eds. 1996. *Structuralist Theory of Science*. New York: Walter de Gruyter.

Barnes, B. 1974. *Scientific Knowledge and Sociological Theory*. London: Routledge & Kegan Paul.

Barwise, J., and J. Etchemendy. 1996. Heterogeneous Logic. In *Logical Reasoning with Diagrams*, ed. G. Allwein and J. Barwise, 179–200. New York: Oxford Univ. Press.

Bechtel, W. 1996. What Knowledge Must Be in the Head in Order to Acquire Knowledge? In *Communicating Meaning: The Evolution and Development of Language*, ed. B. M. Velichkovsky and D. M. Rumbaugh, 45–78. Nahwah, NJ: Lawrence Erlbaum.

Berger, P., and T. Luckmann. 1966. *The Social Construction of Reality*. New York: Doubleday.

Berlin, Brent, and Paul Kay. 1969. *Basic Color Terms: Their Universality and Evolution*. Berkeley: Univ. of California Press.

Bird, Alexander. 1998. *Philosophy of Science*. Montreal: McGill-Queen's Univ. Press.

———. 2000. *Thomas Kuhn*. Princeton, NJ: Princeton Univ. Press.

Blackburn, Simon. 1994. Enchanting Views. In *Reading Putnam*, ed. Peter Clark and Bob Hale, 12–30. Cambridge: Blackwell.

Bloor, David. 1976. *Knowledge and Social Imagery*. London: Routledge & Kegan Paul.

———. 1982. Durkheim and Mauss Revisited: Classification and the Sociology of Knowledge. *Studies in the History and Philosophy of Science* 13:267–97.

Boghossian, Paul A., and J. David Velleman. 1989. Color as a Secondary Property. *Mind* 98:81–103. Reprinted in Byrne and Hilbert (1997a).

Borges, J. L. 1954. *Historia Universal de la Infamia*. Buenos Aires: Emece.

Boyd, Richard. 1984. The Current Status of Scientific Realism. In *Scientific Realism*, ed. J. Leplin, 41–82. Berkeley: Univ. of California Press.

————.1991. Realism, Anti-Foundationism and the Enthusiasm for Natural Kinds. *Philosophical Studies* 61 (1–2): 127–48.

Bridgman. P. W. 1927. *The Logic of Modern Physics.* New York: Macmillan.

Byrne, Alex, and David R. Hilbert, eds. 1997a. *The Philosophy of Color.* Vol. 1 of *Readings on Color.* Cambridge, MA: MIT Press.

————, eds. 1997b. *The Science of Color.* Vol. 2 of *Readings on Color.* Cambridge, MA: MIT Press.

————.2003. Color Realism and Color Science. *Behavioral and Brain Sciences* 26 (1): 3–21.

Callebaut, W. 1993. *Taking the Naturalistic Turn; or, How Real Philosophy of Science Is Done.* Chicago: Univ. of Chicago Press.

Carson, Rachel. 1962. *Silent Spring.* Boston: Houghton Mifflin.

Cartwright, Nancy D. 1983. *How the Laws of Physics Lie.* Oxford: Clarendon Press.

————.1989. *Nature's Capacities and Their Measurement.* Oxford: Oxford Univ. Press.

————.1999. *The Dappled World: A Study of the Boundaries of Science.* Cambridge: Cambridge Univ. Press.

Chandrasekaran, B., J. Glasgow, and N. H. Narayanan, eds. 1995. *Diagrammatic Reasoning: Cognitive and Computational Perspectives.* Cambridge, MA: MIT Press.

Christie, Maureen. 1994. Chemists versus Philosophers Regarding Laws of Nature. *Studies in History and Philosophy of Science* 25:613–29.

Churchland, Paul M. 1989. *A Neurocomputational Perspective: The Nature of Mind and the Structure of Science.* Cambridge, MA: MIT Press.

————.2005. Chimerical Colors: Some Phenomenological Predictions from Cognitive Neuroscience. *Philosophical Psychology* 18 (5):527–60.

Churchland, Paul M., and C. A. Hooker, eds. 1985. *Images of Science.* Chicago: Univ. of Chicago Press.

Clark, Andy. 1997. *Being There: Putting Brain, Body, and World Together Again.* Cambridge, MA: MIT Press.

Collins, Harry M. 1981. Stages in the Empirical Program of Relativism. *Social Studies of Science* 11:3–10.

————.1990. *Artificial Experts: Social Knowledge and Intelligent Machines.* Cambridge, MA: MIT Press.

————.2004. *Gravity's Shadow: The Search for Gravitational Waves.* Chicago: Univ. of Chicago Press.

Collins, Harry M., and S. Yearley. 1992. Epistemological Chicken. In *Science as Practice and Culture,* ed. Andy Pickering, 283–300. Chicago: Univ. of Chicago Press.

Cowan, Ruth Schwartz. 1983. *More Work for Mother: The Ironies of Household Technology from the Open Hearth to the Microwave.* New York: Basic Books.

Cushing, James T. 1994. *Quantum Mechanics: Historical Contingency and the Copenhagen Hegemony.* Chicago: Univ. of Chicago Press.

Daston, Lorraine. 1992. Objectivity and the Escape from Perspective. *Social Studies of Science* 22:597–618.

Daston, Lorraine, and Peter Galison. 1992. The Image of Objectivity. *Representations* 40:81–128.

Devitt, Michael. 1991. *Realism and Truth*. 2nd ed. Princeton, NJ: Princeton Univ. Press.

Drake, Stillman, ed. 1957. *Discoveries and Opinions of Galileo*. Garden City: Doubleday.

Dupré, John. 1993. *The Disorder of Things: Metaphysical Foundations of the Disunity of Science*. Cambridge, MA: Harvard Univ. Press.

Edge, D., and M. Mulkay. 1976. *Astronomy Transformed*. New York: Wiley.

Fairchild, M. D. 1998. *Color Appearance Models*. Reading, MA: Addison-Wesley.

Feyerabend, Paul. K. 1962. Explanation, Reduction, and Empiricism. In *Scientific Explanation, Space, and Time*, ed. H. Feigl and G. Maxwell, Minnesota Studies in the Philosophy of Science 3, 28–97. Minneapolis: Univ. of Minnesota Press.

Forrai, Gábor. 2001. *Reference, Truth and Conceptual Schemes: A Defense of Internal Realism*. Dordrecht: Kluwer.

Friedan, Betty. 1963. *The Feminine Mystique*. New York: Norton.

Galison, P. L. 1997. *Image and Logic: A Material Culture of Microphysics*. Chicago: Univ. of Chicago Press.

Giere, Ronald N. 1985. Constructive Realism. In *Images of Science*, ed. P. M. Churchland and C. A. Hooker, 75–98. Chicago: Univ. of Chicago Press. Repr. in Giere 1999a.

———. 1988. *Explaining Science: A Cognitive Approach*. Chicago: Univ. of Chicago Press.

———. 1994. The Cognitive Structure of Scientific Theories. *Philosophy of Science* 61:276–96. Repr. in Giere 1999a.

———. 1996a. From *Wissenschaftliche Philosophie* to Philosophy of Science. In *Origins of Logical Empiricism*, ed. R. N. Giere and A. Richardson, Minnesota Studies in the Philosophy of Science 16, 335–54. Minneapolis: Univ. of Minnesota Press. Repr. in Giere 1999a.

———. 1996b. Visual Models and Scientific Judgment. In *Picturing Knowledge: Historical and Philosophical Problems Concerning the Use of Art in Science*, ed. B. S. Baigrie, 269–302. Toronto: Univ. of Toronto Press. Repr. in Giere 1999a.

———. 1999a. *Science without Laws*. Chicago: Univ. of Chicago Press.

———. 1999b. Using Models to Represent Reality. In *Model-Based Reasoning in Scientific Discovery*, ed. L. Magnani, N. J. Nersessian, and P. Thagard, 41–58. Dordrecht: Kluwer.

———. 2002a. Distributed Cognition in Epistemic Cultures. *Philosophy of Science* 69:637–44.

———. 2002b. Models as Parts of Distributed Cognitive Systems. In *Model-Based Reasoning: Science, Technology, Values*, ed. Lorenzo Magnani and Nancy Nersessian, 227–42. New York: Kluwer.

———. 2002c. Scientific Cognition as Distributed Cognition. In *The Cognitive Basis of Science*, ed. Peter Carruthers, Stephen Stitch, and Michael Siegal, 285–99. Cambridge: Cambridge Univ. Press.

———. 2003. The Role of Computation in Scientific Cognition. *Journal of Experimental & Theoretical Artificial Intelligence* 15:195–202.

———. 2004a. How Models Are Used to Represent Reality. *Philosophy of Science* 71 (5): 742–52.

———. 2004b. The Problem of Agency in Scientific Distributed Cognitive Systems. *Journal of Cognition and Culture* 4 (3–4): 759–74.

———. 2005. Scientific Realism: Old and New Problems. *Erkenntnis* 63 (2):149–165.

———. Forthcoming. Models, Metaphysics and Methodology. In *Nancy Cartwright's Philosophy of Science*, ed. S. Hartmann and L. Bovens. London: Routledge.

Giere, Ronald N., and Barton Moffatt. 2003. Distributed Cognition: Where the Cognitive and the Social Merge. *Social Studies of Science* 33:301–10.

Gigerenzer, G. 2000. *Adaptive Thinking*. New York: Oxford Univ. Press.

Gilbert, Walter. 1992. A Vision of the Grail. In *The Code of Codes: Scientific and Social Issues in the Human Genome Project*, ed. Daniel J. Kevles and Leroy Hood, 83–97. Cambridge, MA: Harvard Univ. Press.

Glashow, Sheldon. 1992. The Death of Science? In *The End of Science: Attack and Defense*, ed. Richard Q. Elvee, 23–32. Lanham, MD: Univ. Press of America.

Gleason, H. A. 1961. *An Introduction to Descriptive Linguistics*. New York: Holt, Rinehart & Winston.

Golding, Joshua L. 2003. *Rationality and Religious Theism*. Aldershot: Ashgate.

Grant, Peter R. 1986. *Ecology and Evolution of Darwin's Finches*. Princeton, NJ: Princeton Univ. Press.

Greene, Brian. 1999. *The Elegant Universe*. New York: Norton.

Griffiths, Paul E. 1997. *What Emotions Really Are: The Problem of Psychological Categories*. Chicago: Univ. of Chicago Press.

Gross, P. R., and N. Levitt. 1994. *Higher Superstition: The Academic Left and Its Quarrels with Science*. Baltimore: Johns Hopkins Univ. Press.

Gross, P. R., N. Levitt, and M. W. Lewis, eds. 1996. *The Flight from Science and Reason*. New York: New York Academy of Sciences.

Hacking, Ian. 1983. *Representing and Intervening*. Cambridge: Cambridge Univ. Press.

———. 1991. A Tradition of Natural Kinds. *Philosophical Studies* 61 (1–2): 109–26.

———. 1999. *The Social Construction of What?* Cambridge, MA: Harvard Univ. Press.

Hale, B., and C. Wright. 1997. Putnam's Model-Theoretic Argument against Metaphysical Realism. In *A Companion to the Philosophy of Language*, ed. B. Hale and C. Wright, 427–57. Oxford: Blackwell.

Hales, Steven D., and Rex Welshon. 2000. *Nietzsche's Perspectivism*. Urbana: Univ. of Illinois Press.

Hanson, N. R. 1958. *Patterns of Discovery*. Cambridge: Cambridge Univ. Press.

Haraway, Donna J. 1991. Situated Knowledges: The Science Question in Feminism and the Privilege of Partial Perspective. In *Simians, Cyborgs, and Women*, 183–202. New York: Routledge.

Hardin, C. L. 1988. *Color for Philosophers: Unweaving the Rainbow*. Indianapolis: Hackett.

Hilbert, David. 1987. *Color and Color Perception: A Study in Anthropocentric Realism*. Stanford, CA: Center for the Study of Language and Information.

Hoyningen-Huene, Paul, and Howard Sankey, eds. 2001. *Incommensurability and Related Matters*. Boston: Kluwer.

Hull, D. 1988. *Science as a Process: An Evolutionary Account of the Social and Conceptual Development of Science*. Chicago: Univ. of Chicago Press.

Hume, David. 1888. *A Treatise of Human Nature*. Ed. L. A. Selby-Bigge. Oxford: Clarendon.

Humphreys, Paul. 2004. *Extending Ourselves: Computational Science, Empiricism, and Scientific Method*. New York: Oxford Univ. Press.

Hurvich, Leo. 1981. *Color Vision*. Sunderland: Sinauer Associates.

Hutchins, Edwin. 1995. *Cognition in the Wild*. Cambridge, MA: MIT Press.

Ivins, William M. 1973. *On the Rationalization of Sight*. New York: Da Capo.

Jacobs, Gerald H. 1993. The Distribution and Nature of Colour Vision among the Mammals. *Biological Review* 68:413–71.

Johnson, Mark. 1987. *The Body in the Mind: The Bodily Basis of Meaning, Imagination, and Reason*. Chicago: Univ. of Chicago Press.

Kaiser, P. K., and R. M. Boynton. 1996. *Human Color Vision*. Washington, DC: Optical Society of America.

Kevles, Bettyann Holtzman. 1997. *Naked to the Bone: Medical Imaging in the Twentieth Century*. New Brunswick, NJ: Rutgers Univ. Press.

Knorr-Cetina, K. D. 1981. *The Manufacture of Knowledge: An Essay on the Constructivist and Contextual Nature of Science*. Oxford: Pergamon.

———. 1983. The Ethnographic Study of Scientific Work: Towards a Constructivist Interpretation of Science. In *Science Observed*, ed. K. D. Knorr-Cetina and M. Mulkay, 115–40. Hollywood, CA: Sage.

———. 1999. *Epistemic Cultures: How the Sciences Make Knowledge*. Cambridge, MA: Harvard Univ. Press.

Koertge, N., ed. 1998. *A House Built on Sand: Exposing Postmodern Myths about Science*. New York: Oxford Univ. Press.

Kosslyn, S. M. 1994. *Elements of Graph Design*. New York: Freeman.

Kripke, Saul A. 1980. *Naming and Necessity*. Oxford: Basil Blackwell.

Kuhn, Thomas S. 1957. *The Copernican Revolution: Planetary Astronomy in the Development of Western Thought*. Cambridge, MA: Harvard Univ. Press.

———. 1962. *The Structure of Scientific Revolutions*. Chicago: Univ. of Chicago Press.

———. 1970. *The Structure of Scientific Revolutions*, 2nd ed. In *The International Encyclopedia of Unified Science*, vol. 2, ed. Otto Neurath, Rudolf Carnap, and Charles Morris, 53–272. Chicago: Univ. of Chicago Press.

———. 2000. *The Road since Structure: Philosophical Essays, 1970-1993*. Ed. James Conant and John Haugeland. Chicago: Univ. of Chicago Press.

Kuipers, Theo A. F. 2000. *From Instrumentalism to Constructive Realism: On Some Relations between Confirmation, Empirical Progress, and Truth Approximation*. Dordrecht: Kluwer.

Künne, Wolfgang. 2003. *Conceptions of Truth*. Oxford: Oxford Univ. Press.

Labinger, Jay A., and Harry Collins. 2001. *The One Culture: A Conversation about Science*. Chicago: Univ. of Chicago Press.

Lack, David. 1945. *The Galapagos Finches (Geospizinae): A Study in Variation*. San Francisco: California Academy of Sciences.

Lakatos, Imre. 1970. Falsification and the Methodology of Scientific Research Programmes. In *Criticism and the Growth of Knowledge*, ed. I. Lakatos and A. Musgrave, 91–195. Cambridge: Cambridge Univ. Press.

Lakoff, G. 1987. *Women, Fire, and Dangerous Things: What Categories Reveal about the Mind*. Chicago: Univ. of Chicago Press.

Lakoff, G., and Johnson, M. 1980. *Metaphors We Live By*. Chicago: Univ. of Chicago Press.

LaPorte, Joseph. 2004. *Natural Kinds and Conceptual Change*. Cambridge: Cambridge Univ. Press.

Latour, B. 1993. *We Have Never Been Modern*. Cambridge, MA: Harvard Univ. Press.

Latour, B., and S. Woolgar. 1979. *Laboratory Life: The Social Construction of Scientific Facts*. Beverly Hills, CA: Sage. (2nd ed.; Princeton, NJ: Princeton Univ. Press, 1986).

Laudan, Larry. 1977. *Progress and Its Problems*. Berkeley: Univ. of California Press.

———. 1981. The Pseudo-Science of Science? *Philosophy of the Social Sciences* 11:173–98.

Leplin, J., ed. 1984. *Scientific Realism*. Berkeley: Univ. of California Press.

———. 1997. *A Novel Defense of Scientific Realism*. New York: Oxford Univ. Press.

Longino, H. E. 2002. *The Fate of Knowledge*. Princeton, NJ: Princeton Univ. Press.

MacKenzie, Donald A. 1981. *Statistics in Britain: 1865–1930*. Edinburgh: Edinburgh Univ. Press.

Masterman, Margaret. 1970. The Nature of a Paradigm. In *Criticism and the Growth of Knowledge*, ed. Imre Lakatos and Alan Musgrave, 59–90. Cambridge: Cambridge Univ. Press.

McClelland, J. L., D. E. Rumelhart, and the PDP Research Group, eds. 1986. *Parallel Distributed Processing: Explorations in the Microstructure of Cognition*, 2 vols. Cambridge, MA: MIT Press.

Merton, Robert K. 1973. *The Sociology of Science*. Ed. N. Storer. New York: Free Press.

Mollon, J. D. 2000. "Cherries among the Leaves": The Evolutionary Origins of Color Vision. In *Color Perception: Philosophical, Psychological, Artistic, and Computational Perspectives*, ed. Steven Davis, 10–30. New York: Oxford Univ. Press.

Monmonier, Mark. 1991. *How to Lie with Maps*. Chicago: Univ. of Chicago Press.

Morgan, Mary S., and Margaret Morrison, eds. 1999. *Models as Mediators: Perspectives on Natural and Social Science.* Cambridge: Cambridge Univ. Press.

Niiniluoto, I. 1987. *Truthlikeness.* Dordrecht: D. Reidel.

———. 1999. *Critical Scientific Realism.* Oxford: Oxford Univ. Press.

Palmer, Stephen E. 1999. *Vision Science: Photons to Phenomenology.* Cambridge, MA: MIT Press.

Palmquist, Stephen R. 1993. *Kant's System of Perspectives: An Architectonic Interpretation of the Critical Philosophy.* Lanham, MD: Univ. Press of America.

Pickering, Andy. 1984. *Constructing Quarks: A Sociological History of Particle Physics.* Chicago: Univ. of Chicago Press.

———. 1995. *The Mangle of Practice: Time, Agency, and Science.* Chicago: Univ. of Chicago Press.

Psillos, S. 1999. *Scientific Realism: How Science Tracks the Truth.* London: Routledge.

Putnam, Hilary. 1975. *Mind, Language and Reality.* Philosophical Papers 2. Cambridge: Cambridge Univ. Press.

———. 1981. *Reason, Truth and History.* Cambridge: Cambridge Univ. Press.

———. 1999. *The Threefold Cord: Mind, Body, and World.* New York: Columbia Univ. Press.

Ramsey, Jeffrey. 1997. A Philosopher's Perspective on the "Problem" of Molecular Shape. *Synthese* 111:233–51.

Rediehs, Laura J. 1998. *Relational Realism.* Ann Arbor: University Microfilms International.

Resnick, L. B., J. M. Levine, and S. D. Teasley, eds. 1991. *Perspectives on Socially Shared Cognition.* Washington, DC: American Psychological Association.

Ross, Andrew, ed. 1996. *Science Wars.* Durham, NC: Duke Univ. Press.

Rupert, Robert D. 2004. Challenges to the Hypothesis of Extended Cognition. *Journal of Philosophy* 101 (8): 389–428.

Sacks, Oliver W. 1997. *The Island of the Colorblind.* New York: Knopf.

Sargent, Pauline. 1997. *Imaging the Brain, Picturing the Mind: Visual Representation in the Practice of Science.* Ann Arbor: University Microfilms International.

Savage-Rumbaugh, Sue, et al. 1993. *Language Comprehension in Ape and Child.* Chicago: Univ. of Chicago Press.

Sekuler, R., and R. Blake. 1985. *Perception.* New York: Knopf.

Shapere, Dudley. 1964. Review. The Structure of Scientific Revolutions. *Philosophical Review* 73:383–94.

Shapin, S. 1975. Phrenological Knowledge and the Social Structure of Early Nineteenth-Century Edinburgh. *Annals of Science* 32:219–43.

———. 1979. The Politics of Observation: Cerebral Anatomy and Social Interests in the Edinburgh Phrenology Disputes. In *On the Margins of Science: The Social Construction of Rejected Knowledge,* ed. R. Wallis, Sociological Review Monograph 27, 139–78. Keele: Univ. of Keele Press.

———. 1982. History of Science and Its Sociological Reconstructions. *History of Science* 20:157–211.

Simon, H. A. 1978. On the Forms of Mental Representation. In *Perception and Cognition: Issues in the Foundations of Psychology*, ed. C. Wade Savage, Minnesota Studies in the Philosophy of Science 9, 3–18. Minneapolis: Univ. of Minnesota Press.

——.1995. Forward to Chandrasekaran, Glasgow, and Narayanan 1995, x–xiii.

Smith, Glenn S. 2005. Human Color Vision and the Unsaturated Blue Color of the Daytime Sky. *American Journal of Physics* 73 (7): 590–97.

Sneed, J. D. 1971. *The Logical Structure of Mathematical Physics*. Dordrecht: Reidel.

Snyder, John P. 1993. *Flattening the Earth: Two Thousand Years of Map Projections*. Chicago: Univ. of Chicago Press.

Sokal, Alan. 1998. What the *Social Text* Affair Does and Does Not Prove. In *A House Built on Sand*, ed. N. Koertge, 9–22. New York: Oxford Univ. Press.

Solomon, Miriam. 2001. *Social Empiricism*. Cambridge, MA: MIT Press.

Strawson, P. F. 1959. *Individuals*. London: Methuen.

Stroud, Barry. 2000. *The Quest for Reality: Subjectivism and the Metaphysics of Colour*. New York: Oxford Univ. Press.

Suarez, Mauricio. 2003. Scientific Representation: Against Similarity and Isomorphism. *International Studies in the Philosophy of Science*, 17:225–44.

Suchman, Lucy. 1987. *Plans and Situated Actions: The Problem of Human-Machine Communication*. New York: Cambridge Univ. Press.

Suppe, F. 1989. *The Semantic Conception of Theories and Scientific Realism*. Urbana: Univ. of Illinois Press.

Suppes, Patrick. 1957. *Logic*. New York: Van Nostrand.

——.1962. Models of Data. In *Logic, Methodology and the Philosophy of Science: Proceedings of the 1960 International Conference*, ed. E. Nagel, P. Suppes, and A. Tarski, 252–61. Stanford, CA: Stanford Univ. Press.

——.1967. What Is a Scientific Theory? In *Philosophy of Science Today*, ed. S. Morgenbesser, 55–67. New York: Basic Books.

——.1969. *Studies in the Methodology and Foundations of Science: Selected Papers from 1951 to 1969*. Dordrecht: Reidel.

Teller, Paul. 2001. Twilight of the Perfect Model Model. *Erkenntnis* 55:393–415.

——.2004a. How We Dapple the World. *Philosophy of Science* 71 (4): 425–47.

——.2004b. The Law Idealization. *Philosophy of Science* 71 (5): 730–41.

Tennant, N. 1986. The Withering Away of Formal Semantics. *Mind and Language* 1:302–18.

Thelen, Esther, and Linda B. Smith. 1994. *A Dynamic Systems Approach to the Development of Cognition and Action*. Cambridge, MA: MIT Press.

Thompson, Evan. 1995. *Colour Vision: A Study in Cognitive Science and the Philosophy of Perception*. London: Routledge.

Toga, Arthur W., and John C. Mazziotta. 1996. *Brain Mapping: The Methods*. San Diego: Academic Press.

Tomasello, Michael. 1996. The Cultural Roots of Language. In *Communicating Meaning: The Evolution and Development of Language*, ed. B. M. Velichkovsky and D. M. Rumbaugh, 275–307. Nahwah, NJ: Lawrence Erlbaum.

——. 1999. *The Cultural Origins of Human Cognition*. Cambridge, MA: Harvard Univ. Press.

——. 2003. *Constructing a Language: A Usage-Based Theory of Language Acquisition*. Cambridge, MA: Harvard Univ. Press.

Toulmin, S. 1972. *Human Knowledge*. Princeton, NJ: Princeton Univ. Press.

Turkle, Sherry. 1984. *The Second Self: Computers and the Human Spirit*. New York: Simon & Schuster.

Uexkull, Jakob von. 1934. A Stroll through the Worlds of Animals and Men: A Picture Book of Invisible Worlds. In *Instinctive Behavior: The Development of a Modern Concept*, ed. and trans. Claire H. Schiller, 5–82. New York: International Universities Press.

Varela, F. J., E. Thompson, and E. Rosch. 1993. *The Embodied Mind: Cognitive Science and Human Experience*. Cambridge, MA: MIT Press.

van Fraassen, Bas C. 1980. *The Scientific Image*. Oxford: Oxford Univ. Press.

——. 2002. *The Empirical Stance*. New Haven, CT: Yale Univ. Press.

——. 2004. Science as Representation: Flouting the Criteria. *Philosophy of Science* 71 (5): 794–804.

Velichkovsky, B. M., and D. M. Rumbaugh, eds. 1996. *Communicating Meaning: The Evolution and Development of Language*. Nahwah, NJ: Lawrence Erlbaum.

Vygotsky, L. S. 1962. *Thought and Language*. Cambridge, MA: MIT Press.

——. 1978. *Mind in Society: The Development of Higher Psychological Processes*. Cambridge, MA: Harvard Univ. Press.

Wang, D., J. Lee, and H. Hervat. 1995. Reasoning with Diagrammatic Representations. In Chandrasekaran, Glasgow, and Narayanan 1995, 501–34.

Wegener, D. 2002. *The Illusion of Conscious Will*. Cambridge, MA: MIT Press.

Weinberg, S. 1992. *Dreams of a Final Theory: The Scientist's Search for the Ultimate Laws of Nature*. New York: Random House.

——. 2001. Physics and History. In *The One Culture: A Conversation about Science*, ed. Jay A. Labinger and H. M. Collins, 116–27. Chicago: Univ. of Chicago Press.

Weiner, Jonathan. 1995. *The Beak of the Finch: A Story of Evolution in Our Time*. New York: Knopf.

Weinert, F. ed. 1995. *Laws of Nature: Essays on the Philosophical, Scientific and Historical Dimensions*. New York: Walter de Gruyter.

Wertsch, James V. 1985. *Vygotsky and the Social Formation of Mind*. Cambridge, MA: Harvard Univ. Press.

Westphal, J. 1987. *Colour: Some Problems from Wittgenstein*. Oxford: Blackwell.

Williams, Bernard. 1985. *Ethics and the Limits of Philosophy*. Cambridge, MA: Harvard Univ. Press.

Wilson, Margaret D. 1992. History of Philosophy in Philosophy Today; and the Case of the Sensible Qualities. *Philosophical Review* 101 (1): 191–243.

Wolfram, S. 2002. *A New Kind of Science*. Champaign, IL: Wolfram Media.

Woodward, David. 1987. Medieval *Mappaemundi*. In *The History of Cartography*, vol. 1, *Cartography in Prehistoric, Ancient, and Medieval Europe and the Mediterranean*, ed. J. B. Harley and David Woodward, 286–370. Chicago: Univ. of Chicago Press.

Woodward, James. 1989. Data and Phenomena. *Synthese* 79:393–472.

Woolgar, S., ed. 1988. *Knowledge and Reflexivity: New Frontiers in the Sociology of Knowledge*. London: Sage.

Zammito, John H. 2004. *A Nice Derangement of Epistemes: Post-Positivism in the Study of Science from Quine to Latour*. Chicago: Univ. of Chicago Press.

Ziman, John. M. 1968. *Public Knowledge: An Essay Concerning the Social Dimension of Science*. London: Cambridge Univ. Press.

INDEX

A

abstract models, 105–6

achromatopsia, 28, 122n13

Andersen, Hanne, on incommensurability, 130n34

anthroprocentric realism, 121n10

B

Baltzer, W., 128n14

Barwise, Jon, 101

Berger, P., 6

Berlin, Brent, 23

Bird, Alexander, on causal theory of reference, 130n37; on Kuhn, 130n34

Bloor, David, and the "Strong Programme," 7

Boghossian, Paul, 121n8

Borges, J. L., 129n23

Byrne, Alex, 120n1, 122n17, 123n25

C

Callebaut, Werner, 119n23

Caporael, Linnda, 133n3

Carson, Rachel, 117n2

Cartwright, Nancy, 61, 127n5; and "fundamentalism," 118n14

cartography, 75–76

Christie, Maureen, 128n17

chromatic-response function, 19–21; asymmetry of, 121n6

Churchland, Paul: on causal theory of reference, 130n37; on natural kinds, 86; on neuroscience of color vision, 123n19; on superempirical virtues, 131n44

Clark, Andy, 133n8; *Being There*, 114; on extended minds, 108–9; on language and cognitive scaffolding, 115

classical mechanics, 61–62, 69, 70, 87, 94

cognition: collective, 99, 111–12, embedded, 115; embodied, 114

cognitive study of science, 96

Collins, Harry, 119n19; and computers in science, 134n11; and "Empirical Programme of Relativism," 7–8; and "social realism," 119n25

color: constancy of, 19, 25; as dispositional, 121n11; naming, 22–23; as supervening on physical properties, 121n9. *See also* philosophy of color

color objectivism, 25–27; problems for, 30–31

color perspectivism, 31–33; asymmetry of, 32; objectivity of, 33–34

color subjectivism, 23–25; problems for, 30

color vision: comparative, 27–31; evolution of, 29–30; variations in humans, 28–29

complementary wavelengths, 22

COMPTEL, 45–47, 48–49

Compton, Arthur Holly, 45

Compton Gamma Ray Observatory (CGRO), 45–48; as a distributed cognitive system, 107–8

Compton scattering, 45

computer assisted tomography (CAT), 50–51; and computers, 126n9, plate 9; invasiveness of, 126n10

cones, 18
constructivism, 6–11
contingency thesis, 7–10; revisited, 93–95
Copernican model of the universe, 131n41
correspondence rules, 128n9
Cushing, James, 9–10

D
Daston, Lorraine, 125n1
data, models of, 48–49, 68–69
determinism, history of concept, 34
Descartes, Rene, on the nature of color, 121n11
Devitt, Michael, 117n7
diagrams, as tools of reasoning, 101–3
dichromacy: in humans, 28; in other mammals, 29
distributed cognition, 96–99; and human agency, 112–13
distributed cognitive systems: agency in, 108–14; computation in, 107–8; as hybrid systems, 113–14
DNA, model of, 64, 104–5, 106
Dupré, John, on essentialism, 131n40

E
"Empirical Programme of Relativism," 7
Etchemendy, J., 101
external representations, 97–98; pictorial representations as, 104

F
Feyerabend, Paul, on proliferation of theories, 133n49
Fine, Arthur, ix
fitness, 71–72
"Flattening the Earth," 78–80
Forrai, G., 118n11
Friedan, Betty, 117n3
functional magnetic resonance imaging (fMRI), 55–56, plate 12

G
Galison, Peter, 125n1, 126n5; criticism of Pickering, 119n20; on computers in science, 134n11
Galileo: and color subjectivism, 23; and observations of Milky Way, 58, 126n14
generalizations, 67–68
Gigerenzer, Gerd, 134n13
Gilbert, Walter, 118n9
Glashow, Sheldon, 10, 39, 119n22
Golding, Joshua L., 120n29
gravitational lensing, 44
Greene, Brian, 118n9
Gross, P. R., 117n4

H
Hacking, Ian, 118n15, 125n3; and contingency thesis, 8; and natural kinds, 84; on representing, 127n3
Hale, B., 118n11
Hales, Stephen, 117n6
Hardin, C. L., 36; *Color for Philosophers*, 120n1
Haraway, Donna, 120n31
Hering, Ewald, 17, 120n3
heterogeneous inference, 101
Hilbert, David, 120n1, 122n17, 123n25; and "anthroprocentric realism," 121n10
Hooker, Clifford, 131n44
Hoyningen-Huene, Paul, ix; on incommensurability, 130n34
Hubble telescope, 43–45; as distributed cognitive system, 99–100; projected end of, 125n2
hue circle, 17–18
hues, nonspectral and unitary, 18
Hull, David, 133n9; on "species problem," 131n40
Humphreys, Paul, on computers in science, 134n11
Hurvich, Leo, on color interactionism, 31

Hutchins, Edwin: *Cognition in the Wild*, 98–99; on cognition as computation, 133n10; on extended minds, 108, 134n15
hybrid systems, 113–14

I

identification, 62
Imaging Compton Telescope (COMP-TEL), 45–47, plate 6
incommensurability, 82–84
instrumental perspectives, 56–58; compatibility of, 57; overlapping, 57–58
interpretation, 62

J

Jacobs, Gerald, 122nn14–15
Johnson, Mark, 134n13

K

Kant, I., 3, 117n6; and "Copernican Revolution," 120n35
Kay, Paul, 23
Kevles, Bettyann, 127n1
kinds: mechanical, 87–88; objective, 84–86; scientific, 84–88; theoretical, 86–88
Knorr-Cetina, Karin, 7, 118n18, 119n23, 135n19; on extended agency, 109–10
Koertge, N., 117n4
Kosslyn, Stephen, on diagrammatic reasoning, 103
Kripke, Saul, on causal theory of reference, 130n37
Kuhn, Thomas: and incommensurability, 82–84; later writings, 130n35; on living in a different world, 123n23; and paradigms, 82; *The Structure of Scientific Revolutions*, 16
Kuipers, Theo, 128n15

L

Labinger, Jay, 117n4, 119n19

Lakoff, George, 117n7, 134n13; regarding Putnam, 118n11
LaPorte, Joseph, 85, 130n37
Latour, Bruno, 7; and hybrid systems, 113
Laudan, Larry, 119n24
laws of nature, 69–71; and generalizations, 67–68; and principles, 61–62; and theories, 69
Leibniz, G., 3, 117n6
Leplin, J., 117n5, 118n10
Levitt, N., 117n4
Longino, Helen, ix, 117n5

M

magnetic resonance imaging (MRI), 52–55, plate 11; as distributed cognitive system, 100; noninvasiveness of, 126n12
Malin, David, 42–43
maps, 72–75; and models, 76–78
Masterman, Margaret, senses of "paradigm," 130n33
Mercator, Gerardus, 78
Mercator projection, 78–79
Merton, Robert, 120n32
metamerism: for single wavelengths, 21–22; for surface spectral reflectances, 26–27
models: abstract, 105–6; of data, 48–49, 68–69; fit to world, 129n20; and maps, 76–78; as parts of distributed cognitive systems, 100–106; physical, 104–5; and principles, 61–62; representational, 62–63; tests of, 90–92
Moffatt, Barton, 135n24
Monmonier, Mark, 129n27
monochromaticity, 28
Morgan, Mary, 127n5
Morrison, Margaret, 127n5
Moulines, C. U., 128n14

N

naturalism, 11–13; as a methodological stance, 12

natural kinds. *See* kinds
Nersessian, Nancy, 133n3
Newton's laws. *See* classical mechanics
Nietzsche, F., 3, 117n6
Niiniluoto, Ilkka, 117n5, 128n15

O

Oberth, Hermann, 125n2
objective realism, 4–6
opponent-process theory, 18–19
Oriented Scintillation Spectrometer
    Experiment (OSSE), 47–48
overlapping perspectives; instrumen-
    tal, 57–58; theoretical, 92–93

P

Palmer, Stephen, 120n3
Palmquist, Stephen, 117n6
paradigms: and incommensurability,
    82–84; and perspectives, 82
PDP research group, 97–98
perspectival realism, 5–6, 89–92
perspective, in the Renaissance, 14
perspectivism, 13–15
pessimistic induction, 95
Peters projection, 79–80
philosophy of color, 36–37; examples
    of, 123n25; as a science, 38–39
physical models, 104–5
Pickering, Andy, 8–9
pictorial representations, 104
positron emission tomography (PET),
    51–52, plate 10
Pragmatism, 13, 120n30
primary qualities, 37
principles, 61–62
projections of Earth's surface:
    Mercator, 78–79; Peters, 79–80;
    Robinson, 80; Winkel Tripel, 130n32
Putnam, Hilary: and causal theory of
    reference, 130n37; and direct real-
    ism, 35–36; and metaphysical real-
    ism, 4–5; and natural realism, 117n7;

and refutation of metaphysical
    realism, 118n11
Psillos, S., 117n5; about "pessimistic
    induction," 133n50
Ptolemaic model of the universe, 131n41
Pythagorean Theorem, diagrammatic
    proof of, 101–2

Q

quantum theory, 9–10, 34, 52–55, 66, 92

R

realism: anthroprocentric, 121n10;
    about colors, 25–27, 39–40;
    common sense, 4, 117n7; internal,
    4; metaphysical, 4–5; objective,
    4–6; perspectival, 5–6, 88–92;
    scientific, 5
Ramsey, Jeffrey, 128n17
Rediehs, Laura, 118n13
reflexivity: and constructivism, 10–11; and
    perspectivism, 15; reconsidered, 95
relativity: general theory of, 44, 66;
    special theory of, 83, 92
representational models, 62–63
representing, as a four-place relation-
    ship, 59–60
Robinson projection, 80
rods, 120n2
Rupert, Robert, 134n17

S

Sacks, Oliver, 122n13
Sankey, Howard, ix; on incommensu-
    rability, 130n34
Sapir, Edward, 22
Sargent, Pauline, 126n8
Savage, Wade, ix
Savage-Rumbaugh, Sue, 113
science wars, 2, 16, 117n4
secondary qualities, 37; colors as, 37
Shapere, Dudley, review of *Structure*,
    130n33

Shapin, Steven, 7–8
similarity, 63–64; and truth, 64–67
Simon, Herbert, on diagrammatic reasoning, 102
sky, why it is blue, 123n18
Sneed, Joseph, 128n14
"Sociology of Scientific Knowledge" (SSK), 7
Sokal, Alan, 120n36
Solomon, Miriam, "Whig Realism," 133n49
sound, 25, 40
"Strong Programme," 7, 119n27
Stroud, Barry, 39–40, 125nn26–27
Suarez, Mauricio, 128n12
Suchman, Lucy, *Plans and Situated Actions*, 114
Suppe, Fred, ix; on computers in science, 134n11
Suppes, Pat, 68, 128n14
synesthesia, 24

T

Teller, Paul, ix, 61, 128n12, 129n19; on "the perfect model model," 129n18
testing fit of models, 90–92
tetrachromacy: in birds and fishes, 29; in humans, 28–29
theories, 60–61; and laws, 69
theory ladenness of observation, 127n1
Thomasello, Michael, 115
Thompson, Evan, 36, 37, 122n12
tomography, 50
trichromacy, 18, 28
Trifid Nebula, 42–43, plate 4
truth: and commonsense color realism, 39–40; and similarity, 64–67; within a perspective, 81–82
Turkle, Sherry, 134n13

U

Umwelt, 35
uniqueness of the world, as methodological maxim, 34–35

V

van Fraassen, Bas, ix, 119n28, 131n44; statement of scientific realism, 5
Varela, F. J., *The Embodied Mind*, 114
Velleman, J. David, 121n8
visual perspectives: compatibility of, 33–34; objectivity of, 33–34; partiality of, 35–36
von Helmholtz, Hermann, 120n3
von Uexkull, Jakob, 35
Vygotsky, Lev, 135n25

W

Waters, C. Kenneth, ix
wavelength discrimination, 21
Weber, Marcel, ix
Wegener, Daniel, 135n18
Weinberg, Steven, 4–5, 7, 10, 39; *Dreams of a Final Theory*, 117n8, 118n12
Welshon, Rex, 117n6
Whorf, Benjamin Lee, 22
Winkel Tripel projection, 130n32
Wittgenstein, L., on colors, 121n4
Wolfram, S., 134n13
Woodward, James, 68
Woolgar, Steve, 7; and reflexive turn, 119n26
Wright, C., 118n11

Z

Zammito, John, 117n5, 130n36
Ziman, John, and public knowledge, 113